DATE DUE

MAY 12 2004		

GAYLORD

PRINTED IN U.S.A.

Learning from Television
Psychological and Educational Research

Learning from Television
Psychological and
Educational Research

Edited by

MICHAEL J. A. HOWE

Department of Psychology
University of Exeter
Exeter, UK

 ACADEMIC PRESS, INC.

(Harcourt Brace Jovanovich, Publishers)

London   Orlando   San Diego   New York
Toronto   Montreal   Sydney   Tokyo

ACADEMIC PRESS INC. (LONDON) LTD
24/28 Oval Road
London NW1 7DX

United States Edition published by
ACADEMIC PRESS, INC.
Orlando, Florida 32887

British Library Cataloguing in Publication Data
Learning from television.—(Educational Psychology)
 1. Television in education
 I. Howe, Michael   II. Series
 371.3'358   LB1044.7

 ISBN 0-12-357160-X

 LCCCN 82-72665

PRINTED IN THE UNITED STATES OF AMERICA

84 85 86 87   9 8 7 6 5 4 3 2

EDUCATIONAL PSYCHOLOGY

Allen J. Edwards, Series Editor
Department of Psychology, Southwest Missouri State University, Springfield, Missouri

Michael A. J. Howe (ed.). Learning from Television: Psychological and Educational Research

Ursula Kirk (ed.). Neuropsychology of Language, Reading, and Spelling

Judith Worell (ed.). Psychological Development in the Elementary Years

John B. Biggs and Kevin F. Collis. Evaluating the Quality of Learning: The Solo Taxonomy (Structure of the Observed Learning Outcome)

Gilbert R. Austin and Herbert Garber (eds). The Rise and Fall of National Test Scores

Lynne Feagans and Dale C. Farran (eds). The Language of Children Reared in Poverty: Implications for Evaluation and Intervention

Patricia A. Schmuck, W. W. Charters, Jr., and Richard O. Carlson (eds). Educational Policy and Management: Sex Differentials

Phillip S. Strain and Mary Margaret Kerr. Mainstreaming of Children in Schools: Research and Programmatic Issues

Maureen L-Pope and Terence R. Keen. Personal Construct Psychology and Education

Ronald W. Henderson (ed.). Parent–Child Interaction: Theory, Research, and Prospects

W. Ray Rhine (ed.). Making Schools More Effective: New Directions from Follow Through

Herbert J. Klausmeier and Thomas S. Sipple. Learning and Teaching Concepts: A Strategy for Testing Applications of Theory

James H. McMillan (ed.). The Social Psychology of School Learning

M. C. Wittrock (ed.). The Brain and Psychology

Marvin J. Fine (ed.). Handbook on Parent Education

Dale G. Range, James R. Layton, and Darrell L. Roubinek (eds). Aspects of Early Childhood Education: Theory to Research to Practice

Jean Stockard, Patricia A. Schmuck, Ken Kempner, Peg Williams, Sakre K. Edson, and Mary Ann Smith. Sex Equity in Education

James R. Layton. The Psychology of Learning to Read

Thomas E. Jordan. Development in the Preschool Years: Birth to Age Five

Gary D. Phye and Daniel J. Reschly (eds). School Psychology: Perspectives and Issues

Norman Steinaker and M. Robert Bell. The Experimental Taxonomy: A New Approach to Teaching and Learning

J. P. Das, John R. Kirby, and Ronald F. Jarman. Simultaneous and Successive Cognitive Processes

Herbert J. Klausmeier and Patricia S. Allen, Cognitive Development of Children and Youth: A Longitudinal Study

Victor M. Agruso, Jr. Learning in the Later Years: Principles of Educational Gerontology

Thomas R. Kratochwill (ed.). Single Subject Research: Strategies for Evaluating Change

Kay Pomerance Torshen. The Mastery Approach to Competency-Based Education

Harvey Lesser. Television and the Preschool Child. A Psychological Theory of Instruction and Curriculum Development

Donald J. Treffinger, J. Kent Davis, and Richard E. Ripple (eds). Handbook on Teaching Educational Psychology

Harry L. Hom, Jr. and Paul A. Robinson (eds). Psychological Processes in Early Education

J. Nina Lieberman. Playfulness: Its Relationship to Imagination and Creativity

Samuel Ball (ed.). Motivation in Education

Erness Bright Brody and Nathan Brody. Intelligence: Nature, Determinants, and Consequences

António Simões (ed.). The Bilingual Child: Research and Analysis of Existing Educational Themes

Gilbert R. Austin. Early Childhood Education: An International Perspective

Vernon L. Allen (ed.). Children as Teachers: Theory and Research on Tutoring

Joel R. Levin and Vernon L. Allen (eds). Cognitive Learning in Children: Theories and Strategies

Donald E. P. Smith and others. A Technology of Reading and Writing (in 4 volumes)

Vol. 1. Learning to Read and Write: A Task Analysis (by Donald E. P. Smith)
Vol. 2. Criterion-Referenced Tests for Reading and Writing (by Judith M. Smith, Donald E. P. Smith, and James R. Brink)
Vol. 3. The Adaptive Classroom (by Donald E. P. Smith)
Vol. 4. Designing Instructional Tasks (by Judith M. Smith)

Phillip S. Strain, Thomas P. Cooke, and Tony Apolloni. Teaching Exceptional Children: Assessing and Modifying Social Behavior

List of Contributors

ALISON F. ALEXANDER Department of Communication Studies, University of Massachusetts, Amherst, Massachusetts 01003, USA

ANTHONY W. BATES Institute of Educational Technology, The Open University, Walton Hall, Milton Keynes, MK7 6AA, UK

DAN BROWN Department of Communication, University of Evansville, Evansville, Indiana 47702, USA

JENNINGS BRYANT Department of Communication, University of Evansville, Evansville, Indiana 47702, USA

MILTON CHEN Institute for Communication Research, Stanford University, Stanford, California 94305, USA

PETER G. CHRISTENSON Department of Speech Communication, Pennsylvania State University, University Park, Pennsylvania 16802, USA

LEONARD D. ERON Department of Psychology, University of Illinois at Chicago Circle, Chicago, Illinois 60680, USA

MARVIN E. GOLDBERG Faculty of Management, McGill University, 1001 Sherbrooke Street West, Montreal, Quebec, Canada H3A 1G5

GERALD J. GORN Faculty of Commerce and Business Administration, University of British Columbia, British Columbia, Canada

ANNA HOME Television South PLC (TVS), Television Centre, Vinters Park, Maidstone ME14 4NZ, UK

L. ROWELL HUESMANN Department of Psychology, University of Illinois at Chicago Circle, Box 4348, Chicago, Illinois 60680, USA

KEITH W. MIELKE Children's Television Workshop, One Lincoln Plaza, New York, New York 10023, USA

DAVID K. B. NIAS Department of Psychology, Institute of Psychiatry, The Maudsley Hospital, De Crespigny Park, Denmark Hill, London SE5 8AF, UK

GRANT NOBLE Department of Psychology, The University of New England, Armidale, New South Wales 2351, Australia

DONALD F. ROBERTS   Institute for Communication Research, Stanford University, Stanford, California 94305, USA

DOROTHY G. SINGER   Department of Psychology, Family Television Research and Consultation Center, Yale University, 405 Temple Street, New Haven, Connecticut 06511, USA

JEROME L. SINGER   Department of Psychology, Family Television Research and Consultation Center, Yale University, 405 Temple Street, New Haven, Connecticut 06511, USA

Preface

How are we affected by television? The amount of research now being conducted in order to investigate the ways in which children and adults are influenced by television is substantial and growing, although it remains small in relation to the time and importance that modern society gives to television. But detailed accounts of the findings from research are somewhat scattered and inaccessible. Many of the results are described only in technical reports and papers delivered at conferences, or in articles in cross-discipline publications that are not widely read by psychologists and educators. I believe that this book will have the effect of giving a broader group of readers access to important knowledge that is being acquired about some of television's most important effects.

In recent years the questions that researchers have tried to answer concerning the outcomes of viewing television have become more precise, and the methods used in undertaking research have had to be increasingly sophisticated. To most of the simpler queries that can be posed, for instance "Can television educate?" , "Are attitudes, beliefs and knowledge influenced by watching television?", "Does violence on television affect children?", the answer is undoubtedly "Yes", but many other factors determine the extent and manner of the effects. For practical purposes, just establishing the bald answers to these questions provides only a minor advance in our knowledge. It is more important and more interesting to predict who will be influenced and how, and under what circumstances. However, questions like these ones are much harder to answer authoritatively.

It is easy enough to demonstrate that television is an important source of influence upon people's lives. Things are made difficult for the investigating researcher by the fact that the mental processes that govern our knowledge, beliefs, views and attitudes, as well as our actions, are determined by a multiplicity of influences, television being but one. The various factors that affect us do not function singly or even additively, but combine and interact in ways which make it extremely difficult, and in some instances impossible conceptually as well as in practice, to separate the influences of the various different determinants, even when each one is known to be crucial. If we reflect for a moment on how difficult it is to assess the precise effects of any particular ingredient in a culinary recipe, it is not difficult to imagine how exceedingly hard it must be to measure the individual influence of one single factor among the numerous interacting causes and effects that govern people's lives.

Those who investigate the various influences of television share a common interest in adding precision to our knowledge about the effects of the medium. The ten chapters in the present book cover most, but not all, of the major categories of research studies that have been conducted in order to investigate the effects of television upon children and adults. The first three chapters consider the influences of those forms of televis-ion that are deliberately designed to achieve educational goals. Chapter 1, by Jennings Bryant, Alison Alexander and Dan Brown, surveys research that has assessed educational programmes in the United States. In chapter 2 Keith Mielke and Milton Chen consider the design and interpretation processes involved in formative evaluation of a particular education series, *3-2-1 Contact*. The author of chapter 3, Anthony Bates, describes and evaluates the use of television as an integral part of degree-level courses at the Open University in Britain.

Chapters 4 to 8 examine some of the ways in which people are affected by their everyday television viewing. Peter Christenson and Donald Roberts (chapter 4) survey investigations that have examined the various effects of television upon children's social attitudes. Grant Noble, in chapter 5, advances a bold theory suggesting that the particular nature of the television medium and the form of viewers' involvement in the programmes they see combine to influence learning in important and striving ways. The following three chapters describe the effects of viewing specific forms of content. Marvin Goldberg and Gerald Gorn (chapter 6) examine the problems encountered in investigating the effects on children of television advertising. Chapter 7, by Rowell Hues-mann and Leonard Eron, describes research into the effects of televised violence. They emphasise the intervening variables that have to be considered in order to fully understand the influence of violent televis-ion. In chapter 8, David Nias considers evidence about the effects of the equally controversial but less extensively investigated topics of sex and pornography on television, and he discusses some implications of the research findings.

The final two chapters depart slightly from the specific concern with ways in which people learn from what they see on television, and consider the medium from alternative perspectives. Chapter 9, by Anna Home, who is well known in Britain as the creator of a number of excellent children's programmes, writes about children's television and its responsibilities from a producer's viewpoint. In chapter 10, Dorothy and Jerome Singer describe the making and evaluation of a curriculum that has been designed to increase young children's awareness and understanding of the television medium, and help them to be more intelligent "consumers" of the programmes they see.

Together, the ten chapters provide a balanced guide to current research into learning from television. Some of the individual chapters place most emphasis on empirical findings, while others give greater prominence to explanatory theories or to the methodological issues involved in trying to overcome the difficult problems encountered in social science research examining multiply-caused phenomena. Most chapters give special attention to children or young people, but some are concerned, and one exclusively so, with adult learning.

My main hope for this book is that readers will find it genuinely useful. I believe that they will.

February 1983   MICHAEL J. A. HOWE

Contents

Learning from Educational Television Programs

Jennings Bryant, Alison F. Alexander and Dan Brown

Introduction and purpose

During the late 1960s and throughout the 1970s, a number of innovative and popular educational television programs were developed and broadcast. These programs differed dramatically from their predecessors, which had tended to feature lectures from "talking heads". Most of the newcomers utilised the special electronic capabilities of the medium, employed extensive research to refine goals and develop formats, treated specialised topics, sought a well-defined target audience, frequently incorporated social with educational goals, and extensively evaluated the success of their teaching efforts. Although descriptions and critical examinations of each of these programs are available somewhere, many of the best treatments are located in sources such as technical reports to funding agencies or papers presented at professional meetings, so access to information regarding these valuable instructional resources is limited. Moreover, a systematic examination of these several programs in a single source is missing. This chapter seeks to fill this void.

Limitations

Space restrictions typically dictate deletion of desirable content, and the present chapter is no exception. The following criteria were employed to restrict the scope of the investigation: (1) Programs included have been unquestionably identified as educational; predominantly entertainment programs with secondary educational goals have been omitted (e.g. *Captain Kangaroo, Mister Rogers Neighborhood, Over Easy, Fat Albert*). (2) Educational television *systems*, such as are a part of Great Britain's Open University, have not been examined, although the exemplar case is treated elsewhere in this volume. (3) Instructional ("classroom television") offerings have generally been omitted, although a brief section on the Agency for Instructional Television, whose programs tend to defy the traditional educational/instructional dichotomy, has been included. (4) Programs with at least one thorough independent evaluation have been given preferential consideration. Lack of such evaluation makes determining the success of a series in teaching largely a matter of deciphering public relations material. Although the present focus is on findings from summative evaluation, as any examination of "learning" must be, findings from formative research have been incorporated if they provide substantial insight into exactly what viewers received from a program. (5) Innovations and trend-setters have been selected; every program selected for examination has at least one "first" to its credit.

Adhering to these criteria has tended to produce three unintentional "biases" which should be mentioned: (1) The bulk of the programs examined are "children's shows". In general, educating the immature audience has been of priority concern to funding agencies, so proposed educational programs have been funded at levels that have been adequate for "stylish" production and extensive evaluation. (2) Programs by production and/or funding agencies which were responsive to our inquiries or requests for information were reviewed, while information on a few quality programs whose producers or evaluators never sent us promised material remains in "incomplete" files. (3) And, finally, a USA-bias emerged. All non-USA programs that were unearthed through computer searches or other means ended without meeting the previously mentioned criteria. What remains is an impressive array of creative programs with diverse goals and perspectives on tele-education.

Children's television workshop programs

CHILDREN'S TELEVISION WORKSHOP MODEL

The Children's Television Workshop (CTW) has become the dominant force in educational television programing for children. Much of the group's success stems from its conceptual and operational model, which has been described by CTW President Joan Ganz Cooney as a "forced marriage of educational advisors and professional researchers with experienced television producers", a union which became a "howling success" (Lesser, 1974, pp. xv–xvi). For each CTW series, there have been extensive initial planning sessions by producers, researchers, content experts and advisors. The conceptual outcomes of these sessions have then been translated into program segments and pretested on the target audience, frequently for lengthy periods prior to actual program production, so that the producers can actually see just how receptive the audience is to their carefully planned educational messages. This extensive pretesting continues on a day-to-day, segment-by-segment basis during program production, providing immediate and precise feedback instead of the sparse, delayed, and frequently ambiguous feedback which characterises mass communication. In this way, the producer knows what the audience is watching, what it finds particularly appealing, and what is being comprehended (or misinterpreted), so that revisions can take place before the segment is broadcast. This model has been of immense utility to CTW in their goal of teaching through television.

Sesame Street

Sesame Street is a "textbook example" of the potency of combining the methodical, analytical approaches of science with the intuitive, creative methods of television production to evoke planned educational change in young children. The rationale for this fusion of art with science was to create a television program to help prepare children for school, provide for certain needs that "disadvantaged" parents (especially those of the inner city) wanted to have met for their children, and teach young children how to think as well as master normative factual content.

The specific instructional goals for the first season of the series grew from these global objectives and addressed symbolic functions, cognitive processes, and physical and social environments. The symbolic functions included recognition and use of letters, numbers, and geometric forms. The cognitive processes pertained to dealing with events and objects in terms of order, classification, and relationships. Goals relating to the physical environment were directed toward providing general information about natural phenomena and processes, understanding the interdependency relationships that govern natural relationships, and acquiring knowledge of how humans explore and exploit the natural world. The social environment goals pertained to helping children see themselves and others in terms of roles, understanding forms and functions of institutions, seeing situations from other people's points of view, and comprehending the need for social rules such as those rules that protect the interests of justice and fair play (Lesser, 1974). New goals and goals areas were adopted in later seasons, as can be seen from Table 2, which presents the evolving curriculum of the first ten years of *Sesame Street*.

In order to maximise the likelihood of achieving CTW's ambitious first year goals, a staff of in-house formative researchers spent 18 months prior to broadcast pretesting elements and segments designed to be included in *Sesame Street*. Such extensive formative analysis has continued to be a part of this program, as each new segment type is pretested for, at a minimum, attention-getting potential, appeal and comprehensibility. An extremely abbreviated sampler of generalised findings from this formative research on attention suggests that certain programmatic elements are rather consistently attention-getting: animation, pixilation, active animals, rapid pacing, etc. (Reeves, 1970). Basic research on attention by a variety of academicians has tended to substantiate the

Table 1 Profile of Children's Television Workshop programs

Program & funding agency	Date first aired (no. of shows)	Target audience	Goals	Format	Innovations	Areas of evaluation	
						Formative	Summative
Sesame Street US Office of education Ford Foundation Carnegie Corp. US Office of Economic Opportunity Nat. Inst. of Child Health & Human Development Corp. for Public Broadcasting Nat. Foundation of Art & Humanities Markle Foundation & others	Nov. 10, 1969 (over 1600)	3–5-yr-olds	Preschool preparation in symbolic functions, cognitive processes, physical environment, social environment	Magazine, with muppets, documentaries, animation & other special production techniques	CTW Model First mass audience educational program Entertainment-education mix "Slick" production with special effects Preschool literacy curriculum International adoption Formative research methodology	Attention Appeal Comprehension	Viewing Cognitive Attention Learning for achievement Attitudes (to other races) Attitudes to school Cooperation Attitudes to the deaf Effect of parental encouragement to view
The Electric Company US Office of Education Ford Foundation Carnegie Corp. Markle Foundation Mobile Corp. & others	Oct. 25, 1971 (over 700)	Grades 1–4	Reading skills	Magazine, with many humor & attention-getting devices	Reading curriculum Several research techniques Suitability for both home & school viewing Modelling of cognitive (reading) processes on the screen Encouragement of overt verbal responsiveness	Attention Appeal Comprehension	Viewing Cognitive Attitude

Program / Sponsors	Date	Target audience	Goals	Format	Evaluation content	Measures	
Feeling Good Corp. for Public Broadcasting Robert Wood Johnson Foundation Exxon Corp. Aetna Life & Casualty & others	Nov. 20, 1974 (24: 11 h; 13 1/2 h)	Adults (especially young parents, low-income parents & nonregular PTU viewers)	48 behavioral goals/11 health topics	Season A: 1 h, drama, humor & multiple topics Season B: 1/2 h, hosted single topics	Health-care content Prime-time, adult program Evaluation with complementary studies Identified probable viewers for evaluation Interim multiple-sample evaluation Attempt to attract a new (low-income) audience to PBS	Attention Appeal Comprehension	Viewing Cognitive Attitude Health behavior
3-2-1 Contact Nat. Science Foundation US Office of Education United Technologies Corp. Corp. for Public Broadcasting CTW & others	Jan. 7, 1980 (65)	8-12-yr-olds	Promote science & technology, especially to girls & minorities	Magazine, weekly themes with mystery animation adventure serial, documentary, etc.	Proscience content for minorities Extensive evaluation of field film by writers, producers, researchers & scientists Extensive interviews with target audience in pretesting	Needs assessment Format development Appeal Attitude Comprehension Program analyser	

Table 2 *Sesame Street's* evolving goals and curriculum: the first ten years. (From CTW, 1979)

			Curriculum area				
Initiation of curriculum	Prereading & premath skills	Mental Processes	The child & his world	Bilingual & bicultural education	Audiences with special needs	Health practices	
1969–70	Letters Nos 1–10 Geometric forms	Ordering Classification Reasoning Problem solving Relational concepts Perceptual discrimination	Self Roles Differing perspectives Cooperation Fair play The man-made environment				
1970–71	Sight words Nos 1–20 Addition Subtraction	Multiple classification Multiple regrouping Multiple class inclusion Multiple class differentiation Property identification	The mind & its powers Emotions Conflict resolution Audience participation Making inferences Generating explanations & solutions Evaluating explanations & solutions				
1971–72	Verbal blending		Ecology	Spanish culture & art forms Spanish-speaking performers			

Year	Measurement	Sorting by activity	The child & his powers	Spanish sight words	Education for mentally retarded	
1972–73	Measurement	Sorting by activity	The child & his powers Social attitudes	Spanish sight words		
1973–74	More complicated geometric forms		Coping with failure Self-esteem Entering social groups			
1974–75			Creativity: divergent thinking			
1975–76				Taos Indian Pueblo	Education for mentally retarded	
1976–77	Vocabulary development Sight phrases Additional sight words		The role of women	Spanish sight phrases		
1977–78			Prescientific thinking	Hawaii's multicultural & ocean-oriented society		
1978–79			Additional prescientific thinking Relating positively among nonretarded & mentally retarded children	NY City ethnic neighborhood visits	Deafness & signing curriculum	Nutrition Dental care Exercise

Sesame Street formative research findings (e.g. Levin and Anderson, 1976).

The care given to *Sesame Street*'s formative research has been mirrored in its extensive summative evaluation, especially during the first two broadcast years, as learning from the program was assessed by the Educational Testing Service (ETS). The basic questions underlying the summative research were:

What, overall, is the impact of *Sesame Street*?

What are the moderating effects of age, sex, prior achievement level, and SES on this impact?

Do children at home watching *Sesame Street* benefit in comparison with children at home who do not watch it?

Do children in preschool classrooms benefit from watching *Sesame Street* as part of their school curriculum?

Do children from Spanish-speaking homes benefit from *Sesame Street*?

What are the effects of home background conditions on the effects of *Sesame Street*.

In more operational terms, achievement in eight areas of cognitive development stressed by the program was evaluated: body parts, letters, forms, numbers, relational terms, sorting, classification, and puzzles. In later years, as the program has tended to stress prosocial and affective dimensions more frequently, other tests have been derived to refocus the evaluation accordingly.

Ball and Bogatz (1970) reported the ETS summative research from the first year of *Sesame Street* on 943 children, 731 of whom were "disadvantaged". Gain scores (posttest–pretest) on the achievement battery were directly related to amount of viewing for all groups. For similar viewing amounts, younger children achieved larger gains. When the sexes were compared, viewing amount was similar, and heavy-viewing boys equalled the achievement gains of their female counterparts. "Advantaged" children watched more of *Sesame Street* than did any group of "disadvantaged" children and achieved the highest pretest and posttest scores of all groups. Spanish-speaking viewers achieved particularly notable gains in the first season. An attempt to replicate this result in the second season was inconclusive due to a breakdown in the effort to maintain a nonviewing control condition. All groups displayed high attentiveness to characters and production techniques in the show with only slight between-group differences. An "Age Cohorts Study" examined viewing effects within "disadvantaged" at-home viewers and revealed a "clear" effect on gains.

Parents of high learners were more likely than parents of low learners to be educated and to talk to their children about *Sesame Street*. The parents of low learners tended to be less optimistic about their children's schools and teachers, but all high–low learner comparisons had insufficient numbers of subjects for conclusive interpretation. Multivariate analysis of variance (MANOVA) on "disadvantaged" gains in the eight goal areas with independent variables of viewing condition, sex, and viewing location resulted in insignificant independent variable interaction but significant differences in all subgroups on amount of viewing on all dependent measures except Body Parts and Relational Terms. Although significant main effects for home–school viewing and sex were reported, the interaction between these variables was considered more useful. MANOVA with total instead of subtest gains reflected improvement after viewing the show for seven of the eight goal areas. MANOVA with "advantaged" children and with the Age Cohorts Study produced similar results. Ball and Bogatz concluded the existence of an educationally and statistically significant program-viewing impact on most goals areas with no side effects (such as changes in the children's home environments) and equal attraction by the program for both sexes in terms of frequency of viewing, achieving the greatest gains on specific skills.

Bogatz and Ball (1972) analysed the second year of the show to replicate the first year study and examine the new goals added to the curriculum (The New Study). To prevent repeating the first year problem of controls viewing the program, test sites where *Sesame Street* was becoming available for the first time via cable were selected so that noncable residents could become controls. The second year sample contained an increase in "disadvantaged" subjects and assessed achievement on a greater number of goals.

From the New Study (children who had not seen the first year shows), the authors reported that "... it is possible to attribute differences in gains from pretest to posttest to *Sesame Street* viewing and to the encouragement treatment" (Bogatz and Ball, 1972, p. 165). Previous findings of positive goal area effects were supported: of 63 goals taught, 29 were assessed, with "clear impact" in 13, "equivocal impact" in 10, and "no significant impact" in six. Viewers learned basic knowledge and skills from the show but not complex ones. Children who received systematic encouragement to view *Sesame Street* performed as well as other children and were more likely to have their mothers watch and discuss the show with them (76%) than were members of the nonviewing group (60%). Home background elements such as parental expectations for child, parental attitudes toward education, and television viewing habits of child were not changed. Viewing the show positively affected Peabody Picture Vocabulary Test (PPVT) scores used to assess aptitude and readiness for school. No age or sex effects or significant interactions between age and encouragement were found. Low pretest scorers appeared to achieve the largest gains and a low negative correlation existed between socio-economic status (SES) and gains.

Longitudinal assessment of both years in the Follow-up Study revealed that high viewers of *Sesame Street* did not repeat their out-performing of low viewers but did maintain their initial advantage. All follow-up children scored large, consistent gains on the PPVT. The Follow-up Age Cohorts Study indicated better performance on complex goals by repeat viewers than first-time viewers. Because no significant differences were observed in the original goal areas, more goals to benefit continuing viewers were considered as necessary. Despite little previous evidence of attitude change fostered by a television show, repeat viewers displayed significantly more positive attitudes toward school and race of others than did viewers who had seen only the first season and summer of *Sesame Street*.

Teachers who were unaware which pupils had seen the show prior to entering school ranked viewers higher than nonviewers on seven criteria, with attitude towards school and peer relationships reflecting statistically significant differences. The authors did not claim causality but sought to refute charges that *Sesame Street* turns children off to school. Repeated rankings the following spring revealed no differences.

Ball and Bogatz added that the project demonstrated that large-scale field evaluations can produce interpretable, positive results, even when the subjects are preschool children. Television that is entertaining can educate young children in both the cognitive and the attitudinal domains.

Cook and Conner (1976) evaluated the previously presented *Sesame Street* summative research and concluded that the growth observed in the cognitive areas was due to a combination of viewing and encouragement, rather than viewing alone. These authors found the "demonstrated effects of viewing *Sesame Street* without encouragement . . . of limited magnitude and generalizability to date" (p. 164). Liebert (1976) agreed that the data could not establish conclusively that the gains were due to viewing because the groups were not likely to have been equivalent before viewing. Liebert, however, concluded that the "success of *Sesame Street* is that it opened the door for television programs developed to meet children's needs . . ." (p. 171).

In a test of the effects of viewing *Sesame Street* on prosocial development, Paulson (1974) subjected three- and four-year-old disadvantaged, inner-city children to situational tests to ascertain whether such children who watch *Sesame Street* for an hour or more per day are more likely to cooperate than similar children who do not watch the program. The situational tests placed the children in realistic circumstances and determined whether a criterion response was performed. The subjects were selected from day care centers in Seattle, Washington and Portland, Oregon and randomly assigned to the experimental group (viewing of an entire season of *Sesame Street*) or the control group (not viewing *Sesame Street*). Paulson found a clear difference among viewers of the programme and nonviewers and concluded that children who watched the 1971–1972 season of *Sesame Street* learned to cooperate more than the children who did not watch.

Palmer, Chen and Lesser (1976) reported that *Sesame Street* has been broadcast in the "original English language version in more than 40 countries and territories outside the United States including Canada, the Caribbean, Europe, Africa, the Far East, and Australia and New Zealand". Many countries enter coproduction arrangements for *Sesame Street*, where the foreign language adaptations are produced locally and tailored to their individual cultural requirements with significant proportions of new local material. Such arrangements have been made with Spain, Holland, Germany, France, Latin America, Brazil, Kuwait, Sweden, and Israel. Nowhere does CTW merely dub an American version of *Sesame Street* into another language.

The first foreign language versions with local adaptations were in 1972: *Plaza Sésamo* in Mexico City and *Vila Sésamo* in Brazil. *Plaza Sésamo* was the occasion of extensive, carefully controlled examinations of viewers' cognitive achievement. Children four and five years old were randomly assigned to viewing or nonviewing conditions. Tests that were adapted from the ETS measures were given to the 173 children before the initial exposure, after seven weeks, and six months later following broadcasting of 130 programs. Three categories of tests were: (1) "content achievement", in which test item content matched *Plaza Sésamo* segments; (2) "cognitive content", where the test items matched program content with respect to intellectual processes and skills taught by the program; and (3) "independent-cognitive", pertaining to oral comprehension unrelated to the stated goals of the program. Scores by viewers were significantly higher than by nonviewers in all three areas, suggesting benefits extending beyond the specific curriculum goals. The rate of learning apparently increased steadily during the experimental period. Four-year-olds achieved the largest gains, while the smallest gains were made by three-year-olds.

In other countries employing different versions of *Sesame Street*, high proportions of children aged three to five years have elected to watch this innovative program. Resulting measured educational gains have frequently been quite impressive. In Chile, comparisons of achievement between viewing groups and nonviewing groups of preschool children revealed significant differences in favor of viewers of the show on tests of letter and word knowledge, which reflected explicit goals of the series. Such differences were not found, however, on tests of rhythm or auditory discrimination, which were not systematically treated in *Sesame Street*. A test of relational terms showed no significant group differences, but both three- and four-year-old viewers outperformed nonviewing counterparts who were a year older than themselves on tests of letter and word knowledge. In Israel, attention and active participation increased among viewers of the show. Segments tending to discourage attention were heavily didactic, instructional, or verbal. Higher appeal occurred in more visual segments, and cowatching by mothers increased enjoyment and learning. Heavy viewers of the show did better on almost all tests. Children who were new to televison benefited by improved visual processing skills. Similar results were observed in Australia and Mexico. In Germany, mixed results were observed in teaching social attitudes, but Canadian viewers of the show stated more willingness to play with children of other races. In Japan, the show helped older viewers in learning English, and first-time viewers of television in Jamaica characterised as most popular the segments with special effects, animation, and music (Palmer, Chen and Lesser, 1976).

No report on the accomplishments of *Sesame Street* would be complete without mention of its numerous innovations (see Table 1). Indeed, the program is synonymous with innovation in children's educational television. Among its distinctions, it was the first mass audience educational program and still maintains the largest number of young viewers: in 1980–81 *Sesame Street* reached weekly 78% of all 2–5-year-olds within range of its signal, with its signal coverage estimated as 92% of all US television households (A. C. Nielsen Company, 1981). In addition to its phenomenal "reach", *Sesame Street* was the flagship of the "education by entertainment" or "learning can be fun" perspective, the facilitator and product of a rare synthesis of art and science to serve the cause of education, and the pioneer in the use of a child-oriented magazine format with a variety of special production techniques to educate (e.g. animation, pixilation, word balloons, muppets); even its detractors label it "revolutionary" (e.g. Moody, 1980). Its goals – the development of literacy and preliteracy skills for preschoolers – and content were also, at the time of its initiation, unique for television. It was the forerunner of the movement toward extensive formative and summative evaluation, and many of its research techniques were pioneer efforts. Finally, it is unique in the degree of its international adoption – more than 40 countries air the English-language version, and at least eight foreign-language versions serve the children of numerous other nations (Palmer *et al.*, 1976). *Sesame Street* is the most frequently watched, discussed, researched and imitated educational television program in history. Most of the other programs to be discussed are its legacy.

The Electric Company

The Electric Company (TEC) was designed by the Children's Television Workshop to increase reading skills among primary school children grades one through four. The reading skills emphasised were symbol and sound analysis and meaning. Like *Sesame Street*, TEC utilised the CTW model, a magazine format, and a variety of entertaining and attention-grabbing production techniques.

Formative research for TEC provided a number of innovative techniques and critical insights about learning from television, and it will be reviewed in some detail. Pre-production research measured the appeal of characters by eliciting children's responses to photographs in identifying the main characters, describing the role of the character in the program, and ranking characters on appeal by means of sorting the photographs. The children were also asked to tell why they liked/disliked characters in attempts to distinguish between reactions to programs and photographs.

Eye contact with the television screen when TEC segments were shown in the presence of alternative visual stimuli (distractor tests) provided measures of the appeal of particular pieces of experimental production and identified appealing attributes. An innovation of TEC research was the use of time lapse camera instead of a videotape recorder to measure attention to a program. In a comparison of segments falling into high and low attention categories, previous informal selection of optimally appealing attributes was generally confirmed (e.g. chasing themes, slapstick, functionally relevant action, strong rhythm and rhyme, and children on screen). Formative testing was employed to improve the ability of segments to elicit overt verbal (reading) response from viewing children. A coding system was devised to describe all program segments by defining attributes related to appeal (type of humor, theme, characters) and to comprehension (e.g. manner of presenting print, pace, density, or instructional content). Each of these elements was further broken into components (e.g. humor included sight gags, incongruity, puns, whimsy, parody; Rust, 1971a,b).

Comprehension was evaluated using eye movement recording, a stop tape technique, free verbal reports, test analytic procedures, and group observation. Eye movement recording determined whether the viewer was attending relevant aspects of the segments. The method operated by constraining the child's head, beaming a fine point of light onto the cornea, and recording the reflection of the point with a video camera. The record was later superimposed on an image of the material being viewed to provide a composite. Some elements that were found to be positively related to foveal attention were integration of print with action, dynamic presentation of print, and animated format, while negative elements were print in the lower quarter of the screen, distracting action, and rapid presentation of segment material (O'Bryan and Silverman, 1973).

The stop tape technique was the most frequently used method of comprehension testing for TEC. Children were shown individual TEC program elements, and the presentation was interrupted at predetermined points to question the viewers about what had happened, had not happened, or would happen. In early TEC research, story boards were used for this objective. The free verbal report technique was a variation of the stop tape technique in which child responses to completed program segments were solicited and used to identify effective and ineffective segments by criteria of percent of relevant verbalisations. This variation was not considered a useful research tool for TEC because it dealt with formative goals in the short term, while reading mastery is a long-term process; and the intended TEC audience was to be composed primarily of poor readers (Fowles, 1974).

The group observation technique operated under the assumption that comprehension is related to viewer active participation, relating spontaneous verbal responses to the show to program attributes. The production approaches maximising overt reading responses were frequently incorporated into TEC. After the second season, CTW monitored and videotaped visual, verbal, and behavioral classroom responses to programing transmitted from CTW, which remotely controlled the pan, zoom, and focus of the camera in the classroom. Through a transmitter, CTW staff members asked children questions before and after the showing. With the program viewed by the children superimposed in a corner of the CTW monitor, a continuous record of both stimuli and responses was available (Chen, 1972).

Ball and Bogatz (1973) reported the findings of summative research on TEC. The program fostered significant positive effects on reading among children in grades one through four, with the largest improvement among first graders. Overall, gains by viewers over nonviewers were noted for 17 of the 19 reading skills tested, No differences in effects were noted for viewing in black and white *versus* color. No effects were found for amount of home viewing, attitudes of children toward reading or school, attitudes of teachers toward children or reading performance of the children. Parents of TEC viewing first grade children displayed greater confidence that their children were achieving at average levels or above in reading than did parents of first grade nonviewers. Viewers who watched only the first season retained after the second season an advantage in reading achievement over nonviewers, and second season alone proved equally as effective as the first season alone in producing reading gains. Seeing both seasons produced only small increments over viewing a single season.

Independent research by Sproull, Ward and Ward (1976) found an eight week period of viewing TEC too short for preschoolers to learn basic reading skills. However, the children did find the series appealing, were responsive to it, and learned several of the formats of the program. In sum, TEC has been found to be an effective teaching tool for its target age groups.

Feeling Good

Feeling Good was an experimental television series designed by the Children's Television Workshop, using their model, to motivate viewers to take steps to enhance their health and that of their families. The original goals of the program included 11 priority topics: alcohol abuse, cancer, child care, dental care, exercise, health care delivery system, heart disease, high blood pressure, mental health, nutrition, and prenatal care. Forty-eight behavioral goals were incorporated into the series. A major objective, apart from health, was to attract a greater than normal audience to PBS and especially to attract low-income viewers who normally do not watch much public television. The goals of the series were directed toward all adults but targeted especially for young parents and low-income families.

Difficulties in meeting the objectives were encountered early, as funding could not be secured for the entire project, and development occurred in stages, precluding continuity of production staff members and leading to drastic changes, most noticeably in format. The series opened with 11 weekly one-hour programs built around central continuing characters who introduced multiple topics, dealing with the health problems in ways designed as entertaining fare. CTW decided after the sixth program had aired to terminate the format after the eleventh program because of disappointing Nielsen ratings and survey evidence that audiences found the injection of humor to be condescending in the treatment of health subjects. Having remained off the air for about two months, *Feeling Good* returned with 13 weekly half-hour progrmmes hosted by Dick Cavett and treating single topics seriously with no continuing characters. The 48 original behavioral goals were still desired; but the revised emphasis was on information and attitudes, and 33 of the goals were evaluated over 18 of the programs.

The five study evaluation included: (1) a four-city, large sample study of voluntary viewing in a natural setting, (2) a field experiment where low-income and minority audience members were well represented, (3) a series of four national surveys to measure trends of awareness, sources of awareness, viewing of the series, and selected health behaviors, (4) a community monitoring study in which nonreactive institutional measures (e.g. frequent visits to health care centers) could cross validate self reports of series effects, and (5) national audience estimates provided by Nielsen ratings. Each goal was evaluated, and 10 goals yielded partial evidence of effects, and 9 goals yielded no evidence of effects. No clear patterns of effects emerged, perhaps reflecting variations in the amount of treatments given to various goals. Interpretation, however, was complicated by the changing of the format after the evaluations were in progress, and the limited impact of the series appeared to be largely due to lack of public awareness of the existence of the program.

Mielke and Swinehart (1976) cited the innovations of the *Feeling Good* summative evaluation: (1) evaluation with complementary studies, (2) a method of identifying children most likely to view the program, (3) a design comparing viewers of a particular program with people who saw a different program, and (4) taking frequent interim measures, using different samples for each one. More obvious innovations were the targeting of adults and presentation in prime time by CTW.

3-2-1 Contact

CTW developed *3-2-1 Contact* as a series about science and technology for children of ages 8–12. The goals of the series were to help children enjoy science and seek participation in scientific activities; to help children learn different styles of scientific thinking; and to help children (especially girls and minority children) to think of science and technology as fields open to them (CTW News, 1978). Chapter 2 of this volume is devoted to the formative research on this program. The goals were addressed in a magazine format presenting weekly themes based on opposites (e.g. hot and cold), with continuing features (e.g. mystery adventure serial, host characters) and animation, live action film, music, comedy, and special effects.

The 65 program series began in January 1980, offering daily airing of half-hour programs designed to run five days each week for 13 weeks. Beginning in the second season, the program was shown twice daily. The content of the series, produced under the CTW model, was substantially influenced by a formative research program involving more than 10 000 children and 50 studies assessing in the target audience: needs, general characteristics, and factors affecting appeal, comprehension, and role models. Target children were found to be familiar with a wide range of television programs, formats, and characters. The children were articulate in discussing their reactions and preferences regarding television, holding high standards for production values. Attention spans were not so short as for younger audiences, and agreement was high in naming favorite programs. Members of the target group tended to think in concrete terms, not readily making abstractions and having high interest in animal behavior and human anatomy. Stereotyped sex differences were found, with boys being more attracted by science, action adventure programs, outer space settings, and male leading characters. Girls preferred warm human relationships in dramatic fare, brief excerpts from animal films, and female leading characters. The children as a group indicated high appeal of plotted drama over segmented magazine formats in general, strong storylines in documentaries, active visuals over heavy soundtrack emphasis in conveying information and appropriate humor (not "silly" or "babyish"). The children also found explicit connections between segments useful in relating separate elements of programs. In role models, the children especially admired competence in observed child characters, preferring the models of their own age or somewhat older. The children held ambivalent perceptions of scientists and the work of science (Mielke and Chen, 1980).

Although 3-2-1 Contact has not received formal summative evaluation, a number of indices indicate that it achieved its first year goals, educational and otherwise. For example, Nielsen rating estimates indicate that more than 23 000 000 viewers tuned to the program at least once over its 13-week initial broadcast period (including 73 million tune-ins among households with children under the age of 18); 300 000 teacher guides are in circulation; numerous curriculum committees have given the show high praise; and stacks of letters to CTW have documented viewer interest in and learning from the program (Mielke, 1981).

OTHER CTW PROGRAMS AND ACTIVITIES

Other CTW programs, such as *The Best of Families* and *The Lion, the Witch and the Wardrobe* achieved critical acclaim and strong viewership; however, they will not be reviewed here because their primary goals appear to be dramatic, with education included as a strong subsidiary goal. Other CTW activities, such as the *Sesame Place* theme parks and CTW toys, games, and records are also noteworthy for their education-related achievements but must be left for review in alternative sources.

Other programs

GENERALISATIONS

CTW programs were singled out because of their unique posture in educational programing. A host of other educational or production agencies, including a sizable number of public television stations, have provided effective and innovative educational television programs. Some like *Freestyle* and *Vegetable Soup*, have received research attention roughly equivalent to that devoted to a CTW program. Others have received only formative or only summative evaluation, thereby reducing the number of educationally relevant findings. Frequently, full research evaluation is lacking because a single agency must be in existence for sufficient time to complete planning, execution, and follow-up, and many educational program producers have been able to secure funding only for limited periods. Nonetheless, each program to be considered has made a major contribution to teaching via television.

PROGRAM ANALYSIS

Freestyle

Freestyle was a public television series produced in 1978 by the Television Career Awareness Project in cooperation with several major organisations, including public television station KCET, Community TV of Southern California, and the Annenberg School of Communications at the University of Southern California. Directed to 9–12-year-old children, the goals of the series were to combat sex-role stereotypes and expand career awareness by treating topics pertaining to childhood preoccupational activities, childhood behavioral skills, and adult work and family roles. The treatment was accomplished in dramatic stories featuring an ethnically diverse cast (Johnston, Ettema and Davidson, 1980).

Formative research by LaRose (1978) documented the existence of stereotyped attitudes, beliefs, and interests among members of the target audience, with influential others best predicting certain stereotyped behaviors. Segments of *Freestyle* were better liked and understood by girls than boys. Black and Chicano children liked the material more but

Table 3 Profile of four children's educational television programs

Program, funding agency & production agency	Date aired (no. of shows)	Target audience	Goals	Format	Innovations	Areas of evaluation Formative	Summative
Freestyle Nat. Inst. of Education KCET-TV, Los Angeles, California	1978 (13 ½ h)	9-12-yr-olds	Combat sex & ethnic stereotypes	Dramatic episodes, 2 per show	Theoretical model of stereotyping & change Correlated subject personality with formative evaluation Focus on modelling intentions Connecting talents' & viewers' significant others Emphasis on attitudes in summative evaluation Metric scaling in evaluation Longitudinal evaluation	Appeal Comprehension Modelling intentions	Beliefs Attitudes Interest
Carrascolendas US Office of Education KLRN-TV, Austin, Texas	1970 (regionally) 1972 (nationally) (60)	6-8-yr-olds Spanish speaking	Improve speaking of Spanish Promote pride in Spanish culture	Magazine, with puppets, dramatic skits, songs, animation	Spanish-language curriculum Bilingual show & evaluation Abstract questions Narrow target audience Nonverbal affective responses Devaluated	Attention Comprehension Appeal Cognitive Verbal modelling	Cognitive Attitude Affective Reported behavior
Around the Bend Appalachia Educational Laboratory Nat. Inst. of Education WSAZ-TV, Huntington, West Virginia	1968 (172)	3-5-yr-olds Appalachians	Televised kindergarten Develop language, cognition, psycho-motor, social skills	Multiple segments, continuing heroine, some animation film, puppets, etc.	Use of normed measures Early developer of overt observational measures Complementary aide: home visits, mobile classroom Model intentions Early social-skills curriculum		Cognitive Behavior Appeal Attitudes Standardised perceptual tests
Big Blue Marble Internat. Telephone & Telegraph BBM Productions	1974 (171)	Grades 4-6	Encourage international awareness	Magazine, with animation, mini-documentaries, pen-pals	Cultural value emphasis Semantic differential evaluation Pen-pal feature	Affect Comprehension	Viewing Appeal Attitude Comprehension

understood it less than did white children. This research indicated a need of concrete and explicit treatments, minimising abstractions (LaRose, 1978; Williams, 1978a,b). Williams, LaRose and Frost (1981) reported that congruence of sex of viewer and program character was the most significant predictor of modelling intention. Other consistent predictors were physical attractiveness of the character and the belief that significant others of the viewer would support the modelling.

Johnston and Ettema (1982) have recently reported on the *Freestyle* summative research program: they note that intensive classroom viewing of *Freestyle* was found to produce more positive attitudes in children toward people who behave nontraditionally, such as (*a*) girls who participate in athletics or mechanics, display leadership qualities and independence; (*b*) boys who show nurturing behavior; and (*c*) adults of both genders pursuing nontraditional careers. In other words, viewing *Freestyle* was considered to have changed the norms held by children about sex-roles.

Not only were attitudes toward people engaging in these nontraditional activities influenced, but beliefs about how well the activities could be performed were apparently altered. Girls came to be seen as more competent mechanics and leaders, for example, and boys were considered as competent nurturers.

Viewing the program was least effective in promoting children's interest in non-traditional activities, although subgroups of children were reported as having changed interests a great deal. Attitudes and beliefs were more strongly influenced than were interests.

Johnston and Ettema (1982) found the persistence of effects nine months after the end of the experimental viewing to be strong. They noted "no decay at all" in some areas but "almost total decay" in a few others, with the typical decline in the range of 25–40% of the original effect. An educationally significant net effect resulted in most cases.

These findings were obtained with children viewing *Freestyle* and then participating in classroom discussion with peers. Johnston (1982) summarised two experiments designed to measure the effects of mere viewing. He found the results to indicate "some highly valued changes" but concluded that the prediction of which changes is difficult.

Several noteworthy innovations emerged from the *Freestyle* project, especially involving the formative and summative research efforts: a theoretical model of sex-role stereotyping development and change (Williams *et al.*, 1981); a procedure correlating subject personality measures with formative evaluation measures; a procedure focusing on modelling intentions; a connecting of the significant others of viewers with the significant others of program talent; an emphasis on viewer attitudes in summative evaluation; a scaling evaluation procedure; and a nine month follow-up to test persistence of effects.

Carrascolendas

The United States Office of Education (Title VII) funded *Carrascolendas*, which was produced by KLRN-TV, Austin, Texas. Sixty programs were distributed regionally in 1970 and nationally in 1972 and directed to six- to eight-year-old Spanish-speaking children. The bilingual show was designed to improve the speaking of Spanish and to promote pride in Spanish culture through a format using puppets, dramatic skits, songs, and animation.

Formative research on *Carrascolendas* found high eye contact across the programs, with frequent smiling and laughing; more eye contact for girls than for boys; positively related eye contact with (1) level of perceptual cognitive development and (2) comprehension and recall; negatively related eye contact with verbalisation and modelling; few imitations by viewers and few program-related verbalisations by viewers; and comprehension and recall of two-thirds of the program content. Significant intergroup differences were found for the variables of eye contact, verbal modelling, and smiles and laughter (Laosa, 1974).

The findings of summative research included an increase in knowledge of Spanish culture and history (Williams, van Wart and Stanford, 1973; van Wart, 1974), increased pride in Spanish culture (Williams *et al.*, 1973; van Wart, 1974), and increased use of Spanish by children reported by teachers and parents (Williams and Natalicio, 1972; Williams *et al.*, 1973; van Wart, 1974).

Carrascolendas provided innovations in (1) testing with bilingual questions that were "far beyond" typical acquisition questions in abstraction level and (2) evaluation of nonverbal responses by smiles and laughter. It was also one of the first educational programs directed toward a narrowly-defined target audience.

Around the Bend

Around the Bend, a project of the Appalachia Educational Laboratory, was produced by WSAZ-TV in Huntington, West Virginia. This pioneer program aired the first of 172 episodes in 1968. The target audience was three- to five-year-old residents of Appalachia, and the program was one component in an integrated system including daily half-hour telecasts, paraprofessional home visitors, and a self-contained mobile classroom. The goals of the system were to provide an alternative to conventional kindergarten for Appalachia (where few kindergartens were available) residents and to prepare the target children in such areas as language, cognition, psycho-motor skills, and social skills. The program was in black and white in a format focusing on a young woman at home doing things which could be imitated by viewers. Major techniques included animation, film, puppets, stories, music, models, and demonstrations.

A variety of evaluative reports indicate positive receptions from children and parents, adjustment in the program during the second year on the basis of findings, higher achievement by program viewers, after viewing than controls, and increased social skills after exposure to mobile classroom experiences (Pena and Miller, 1971; Cagno and Shively, 1973).

To summarise a variety of studies (Bertram, Pena and Hines, 1971; Pena and Miller, 1971; Hines, 1973; Cagno and Shively, 1973), researchers found the viewing of *Around the Bend* interesting and appealing (especially with short segments and a variety of production techniques) to the target children; effective in generating spontaneous viewer responses, curiosity, positive attitudes, better verbal expression, and more gramatically correct expression; and significantly greater eye and motor coordination and spatial perceptions. Second year programs were found to be more effective in eliciting enthusiasm and responses. A significant viewing effect was observed for the combination of viewing, home visits by paraprofessionals, and mobile classroom, over viewing only and controls.

Big Blue Marble

The *Big Blue Marble* was funded by International Telephone and Telegraph and produced by Big Blue Marble Productions in 1974. The goal of encouraging international awareness was directed toward viewers in grades four through six. The series employed a magazine format featuring segments such as animated folk tales and mini-documentaries from around the world. *Big Blue Marble* was shown all over the world.

Educators were consulted on thematic focus, format, and evaluation of decisions made and were involved in the early planning of goals and scripts and in reviewing the pilot. Teacher and student reaction to four pilots revealed positive effect and favorite segments (i.e. appeal). Any student difficulties with the pilots were noted for revision.

Summative results indicated that viewers saw more similarity among people from all parts of the world than nonviewers perceived. An increase was observed in viewers' perceptions of others' well-being. While viewers from all three grades reflected a decrease in ethnocentrism, fourth and sixth graders were less sure that the non-Americans would prefer the USA (Roberts, Herold, Hornsby, King, Sterne, Whitely and Silverman, 1974).

The program was innovative in providing foreign pen-pal addresses (7000 children requested one by mail), in addressing cultural (not social) values, and in using semantic differentials as a measurement device.

Vegetable Soup

The Bureau of Mass Communication of the New York State Education Department created *Vegetable Soup* to combat ethnic stereotypes and promote racial understanding among elementary school age children. Employing a magazine format and a variety of production techniques, the series advocates cultural diversity and positive attitudes toward other ethnic groups. The theme song proclaims allegorically: "It takes all kinds of vegetables to make a vegetable soup".

Each half-hour *Vegetable Soup* program is comprised of a number of different segments, although continuing characters and plots provide continuity between programs. For example, a segment from *Outerscope I* (an adventure series in which puppet children of different ethnic backgrounds travel through space encountering prejudice-laden situations) always ends on a "cliff-hanger" which is serialised into the next program. Other segment genres include *People's jobs* (mini-documentaries on minority men and women in various occupations), *Crafts, recipes and games* (animated "how to" segments encouraging cultural diversity by introducing new foods, games and crafts), and *Children's questions about race* (e.g. "Why don't all people smell the same?").

Vegetable Soup received both formative and summative evaluation (e.g. Mays, 1977; Mays, Henderson, Seidman and Steiner, 1975). Formative evaluation provided an initial assessment of target audience members' attitudes toward ethnicities in addition to testing specific materials for appeal, attention, and comprehension. Summative research on the effects of exposure to 16 programs incorporated a posttest-only/control group design with sex, age (6–8, 8–10), and racial ethnic group (black, white, Asian, Puerto Rican, Chicano) as additional factors.

The findings of the summative research covered attitudinal and/or affective dimensions. An investigation of whether program viewers demonstrated more positive identification and awareness of their own ethnic group asked children to select photographs of children most resembling the viewers themselves and of who could be in their families. No statistically significant differences emerged between viewers of *Vegetable Soup* and members of a control group, although more selections by the experimental group were racially identified.

The question of whether the program viewers developed greater feelings of acceptance for others new to their group was examined by having subjects project their reactions to a photograph depicting the arrival for the first time of a new child in school. A significant interaction between race and program exposure apparently resulted from white children, who scored significantly higher in their feelings of acceptance for new children of diverse races than did white children in the control group. The effect for program exposure for all children, however, was not significant.

Sorting 24 photographs of individuals of different races was used to test whether *Vegetable Soup* viewers were more friendly to others. The experimental group members exhibited a significantly greater level of friendliness toward other children than did the control group members.

Sketches portraying occupations were matched by the children to the 24 photographs of children of different ethnic groups to assess the tendency of experimental and control group members to stereotype what people could do occupationally on the basis of sex and race. No significant differences emerged.

Table 4 Profile of five children's educational television programs

Program, funding agency & production agency	Date aired (no. of shows)	Target audience	Goals	Format	Innovations	Areas of evaluation Formative	Summative
Vegetable Soup NY State Education Dept.	1975 (78)	6–10-yr-olds	Combat ethnic stereotypes Promote racial understanding	Magazine, with various production techniques	Prosocial content Nonverbal appeal measures	Attention Attitudes Appeal Comprehension	Attitude Affective
Infinity Factory US Dept. of Education Carnegie Corp. Markle Foundation Nat. Science Foundation Sloan Foundation	1976 (52)	8–11-yr-olds	Math skills Prosocial problem solving, especially for blacks & Latinos	Magazine, polyethnic casts, various locations	Math curriculum Assessed attitudes to math TV program Combining prosocial with math		Attention Appeal Interest Attitude Cognitive
As We See It US Dept. of Education WTTW-TV, Chicago, Illinois	1977 (52)	High-school students	Facilitate high-school desegregation	Documentaries, docudramas, fantasy-satires	Desegregation content Codeveloped & co-produced by target audience	Attention Appeal Comprehension Attitude	
Caboodle US Dept. of Education KLRN-TV, Austin, Texas	1978 (13)	5–8-yr-olds	Art & humanities appreciation and education	Electronic field trips	Art/humanities content Questions about movement & rhythm	Attention Empathy Appeal Comprehension	
Khan Du US Office of Career Education KLRN-TV, Austin, Texas	1978 (4)	6–11-yr-olds	Handicap awareness, attitude & self-image	Documentary & drama with animated heroine	Career awareness for handicapped Self-esteem assessment Evaluation by therapists	Attention Appeal Attitude	Self-esteem Cognitive Attitude

Although *Vegetable Soup* failed to achieve demonstrable educational success on a number of dimensions, it has been extremely popular and has received considerable critical acclaim from educators.

Infinity Factory

Infinity Factory was developed by the Education Development Center to facilitate 8–11-year-old children's understanding of basic mathematics skills useful to everyday life (e.g. decimal number system, measurement, scaling) and to provide models of creative and prosocial problem-solving, especially for black and Latino children, who were a primary target audience. The series used a magazine format, a multiethnic cast, and a variety of urban locations; each program featured one main theme involving two or three basic skills or concepts.

A test of the effects of viewing eight *Infinity Factory* programs was conducted on over 1000 students in conjunction with a trial broadcast season in 1976. The evaluation included assessments of attention, appeal, and comprehension, as well as mathematics learning and the effect on attitudes toward mathematics.

Overall attention to the programs was very high (91.3%). They were also rated as very appealing, especially by black and Latino children. Comprehension of the story line averaged 77% of absolute, with no differences between whites, blacks, and Latinos. In terms of acquisition of information regarding mathematics, all groups showed significant mathematics performance gains at posttest, although white children showed significantly greater gains than blacks or Latinos. All groups of viewers were significantly more positive than nonviewers about mathematics and about mathematics programs on television (Harvey, Quiroga, Crane and Bottoms, 1976).

As We See It

Public television station WTTW-Chicago was funded by the Emergency School Aid Act (ESAA) to produce two series of programs (26 programs each – *As We See It* – I and II) to improve high school students' understanding of social issues relating to desegregation among racially and ethnically diverse groups (Hsia and Strand, 1977; Strand, 1979). The series was codeveloped and coproduced by teams of students from the target population. It featured documentaries, docudramas, and humorous fantasy satires of high school desegregation experiences. Although the series apparently received no summative evaluation, pilot programs for both the 1977 and 1979 series were extensively tested, providing some summative-type evidence regarding learning from the programs.

The pilot testing involved nationwide quota samples of almost 2500 high school students. The pilot testing included observation of audience members' viewing behavior (attention, appeal), postviewing student questionnaires (information acquisition, comprehension, talent evaluation, segment appeal), small-group semi-structured discussion (appeal, attitude change, general reactions), teacher interviews (appeal, pedagogical utility), and one-month delayed follow-up small-groups student discussion (information retention, affect and attitude change retained). The *results* were: (1) *attention and appeal*: Over 85% of the students were rated as generally attentive and enthusiastic. Nearly 80% of the viewers reported that they would like to see the remainder of the shows in the series. (2) *Information acquisition and comprehension*: Viewers performed significantly better than nonviewers (control group) on 7 of 8 test items designed to determine how much they knew about desegregation. (3) *Attitude toward desegregation*: Viewers showed significantly more favorable attitudes toward desegregation than nonviewers and indicated greater acceptance of members of other ethnic groups as a friend or coworker.

Although the lack of summative evaluation severely qualifies the conclusions that can be drawn on learning from *As We See It*, the comprehensive pilot-testing suggests that this series was educationally effective. It should also be noted that *As We See It* was innovative in the extent to which the target audience contributed to the development and production of the programmes.

Caboodle

Caboodle, a 1978 KLRN-TV (Austin, Texas) production funded by the Office of Education (US Department of Health, Education and Welfare), was directed to children in kindergarten through grade three. The 13 programs in the series were intended to introduce concepts from the arts and humanities by leading the viewers on an electronic field trip. Topics include painting, sculpture, architecture, music, dance, and drama.

Formative research in a laboratory setting revealed that children in the target group could have empathy for an appealing adult character. When child characters, especially of varied age and ethnic background, appeared in test segments, viewer attention was high. Long dialogues brought low viewer attention. Field tests assessed student opinions of likeability and preference for scenes and actors, interest level, and comprehension, leading to revisions in the program (Southwest Texas Public Broadcasting Council, 1979).

Innovations of the series were in content and questions relating to movements and rhythm.

Khan-Du

Khan-Du, a 1978 production of KLRN-TV (Austin, Texas), was funded by the Office of Career Education (US Department of Health, Education and Welfare). The four program series was directed to 6–11-year-old children and designed to encourage a positive self-image in the handicapped, to develop community awareness of the problems of the handicapped, and encourage a more positive attitude toward the handicapped by nonhandicapped people. The series featured an animated heroine, documentaries, and dramas.

Formative research with 80 students (40 handicapped) revealed that students liked the program, lost attention during documentaries, liked the heroine, and agreed that handicapped people "can do" for themselves.

Summative findings included no significant increase in self-esteem of handicapped children as measured on a scale, although adults surveyed reported that the program should result in higher self-esteem for members of the target audience. Higher awareness was noted of the functioning abilities of the handicapped, especially in nonhandicapped students.

One of the innovations of the programs was utilising students from a school for the blind to test television programs (Boyd, 1979).

The Agency for Instructional Television

The Agency for Instructional Television (AIT) has pioneered in innovative instructional television programing, including programs which employ affective goals as primary criteria for evaluation. The agency has typically employed attention-getting, entertainment-oriented approaches in developing a number of low-budget but effective programs for children. The productions are frequently mini-dramas featuring school children and adults who are not professional actors. A partial list of AIT programs includes *Inside/Out* (30 brief plays directed to 8–10-year-old children dramatising crises in everyday living), *Ripples* (36 plays about human relationships), *Images and Things* (30 programs for 10–13-year-old children about perceiving beauty in the world through art), *Bread and Butterflies* (a career development series for 9–12-year-old children), *Thinkabout* (a series designed to teach 10–12-year-old children the skills essential to learning), *Self Incorporated* (designed to probe the identity problems of early adolescents), and *Trade-offs* (a series for 9–13-year-old children designed to teach the basics of economics).

AIT is also a pioneer in creative financing for educational/instructional programing. Rather than relying on government or corporate grants, they are a nonprofit program cooperative. After AIT determines the market for a particular proposed series, it obtains production money from charter subscribers, who receive free unlimited use of the programs, which may also be rented from AIT by other users. This is a fiscal model which has substantial promise during times of limited resources for educational television.

Summary and conclusions

The dominant generalisation to be drawn from analysing these outstanding programs is that viewers can and do learn from educational television. While none of the programs treated in this chapter were completely successful in developing an educational program directed toward adults, the evidence of success with children as the target audience appears to be overwhelming. The experience with children suggests, furthermore, that similar achievements in adult educational programing await only a comprehensive endeavor with the right subject matter. The production of educational programs which are appealing, comprehensible, and attention-getting and produce desired affective, cognitive, attitudinal (expressed), and behavioral changes has been documented. Some specific examples of these outcomes are: improved reading skills, mathematics skills, PPVT scores (estimate IQ and readiness for school), visual processing skills, spatial perceptual skills, knowledge of health and human anatomy, attitudes toward entering school, attitudes toward people of different races, social functioning, cultural awareness, self-esteem, cultural pride, and much more.

Television had a 20-plus year history without such educational achievement. In that sense, the programs described here were a dramatic departure from the status quo. The successful teaching by these programs was largely due to careful planning and evaluation. The outcomes were not the result of chance nor merely of the blend of creativity with opportunity. Instead, the combined efforts of educators and television producers produced the desired ends; frequently, for different audiences, for both genders, and covering many topics. Formative research was used to refine goals and formulate strategies for accomplishing the goals. Summative research evaluated how well the finished products performed so that future efforts could benefit from the experience.

Given considerable educational success, particularly on the part of some programs, a search for the "magic formula" might appear to be in order. The programs that appeared to be most successful in achieving their goals began with extensively-examined, well-defined goals packaged in well-produced, entertaining programs appropriately aimed for the target audience. For programs that achieved the most successful educational results, supplemental components accompanied the telecasts: classroom use of programs and home visits by supporting staff members were useful, but an essential "supplemental" ingredient was effective promotion of the program. And, finally, the most difficult aspect of the formula for the educator/producer to provide, cooperative viewing and discussion at home was a consistently important factor in learning from educational television.

Impressive as the accomplishments of the 1960s and 1970s have been in the area of educational television, several problems remain. Educational gaps between "advantaged" and "disadvantaged" populations have not vanished in the application of television to teaching. The entertaining features of successful programs attract members of both groups, perpetuating the discrepancies between them. The gains achieved by viewers of the programs under discussion were accomplished at considerable financial expense for planning, production, and evaluation; yet government funds for social and educational improvement projects appear to be diminishing in the countries which have been the heaviest producers of educational television programs. Alternative funding must be found to make these ventures viable for the future. Perhaps the financial model used by the Agency for Instructional Television need to be more extensively employed in educational television programing.

Whatever problems still exist, however, do not alter the observed successes of television as teacher. In fact, the development of the new specialised media holds renewed promise for teaching via television.

Acknowledgements

The authors wish to express their gratitude to Leslie Winston for her assistance in preparing this chapter. The following scholars were most cooperative in providing primary resource materials and/or editing an earlier draft of this chapter: Edward L. Palmer, Keith W. Mielke, Gerald S. Lesser, James S. Ettema, and Jerome Johnston. Their generous contributions of time, talent, and knowledge were invaluable to the authors.

References

Ball, S. and Bogatz, G. A. (1970). *The First Year of* Sesame Street: *An Evaluation*. Educational Testing Service, Princeton, New Jersey

Ball, S. and Bogatz, G. A. (1973). "Reading with Television: An Evaluation of *The Electric Company*". A Report to the Children's Television Workshop. Educational Testing Service, Princeton, New Jersey

Bertram, C. L., Pena, D. and Hines, B. (1971). "Evaluation Report: Early Childhood Education Program, 1969–70 Field Test. Summary Report". Appalachia Educational Laboratory, Charleston, West Virginia

Bogatz, G. A. and Ball, S. (1972). *The Second Year of* Sesame Street: *A Continuing Evaluation*, Vols I and II. Educational Testing Service, Princeton, New Jersey

Boyd, C. (1979). *Final Project Performance Report*. Office of Career Education, US Department of Health, Education and Welfare, Washington

Cagno, D. and Shively, J. (1973). "Children's Reactions to Segments of a Children's Television Series". Technical Report, No. 34. Appalachia Educational Laboratory, Charleston, West Virginia

Chen, M. (1972). *Verbal Response to* The Electric Company: *Qualities of Program Material and the Viewing Conditions Which Affect Verbalization*. Children's Television Workshop, New York

Children's Television Workshop. (1978). "Final Report on a Television Reading Series". Children's Television Workshop, New York

Children's Television Workshop. (1979). Corporate Review. *Sesame Street*, 10th Season, pp. 10–11. Children's Television Workshop, New York

Children's Television Workshop News. (1978). *Children's Television Workshop Plans Major New Educational Series for 1980*. Children's Television Workshop, New York

Cook, T. D. and Conner, R. F. (1976). *Journal of Communication* 26(2), 155–164

Diaz-Guerrero, R., Reyes-Lagunes, I., Witzke, D. B. and Holtzman, W. H. (1976). *Journal of Communication* 26(2), 145–154.

Fowles, B. R. (1974). "A Pilot Study of Verbal Report in Formative Research in Television". Doctoral dissertation, Yeshiva University, New York

Harvey, F. A., Quiroga, B., Crane, V. and Bottoms, C. L. (1976). *Evaluation of Eight Infinity Factory Programs*. Education Development Center, Newton, Massachusetts

Hines, B. (1973). "Attainment of Cognitive Objectives". Technical Report, No. 3. Appalachia Educational Laboratory, Charleston, West Virginia

Hsia, J. and Strand, T. (1977). *TCR '77: Formative Evaluation*. Office of Education, US Department of Health, Education and Welfare, Washington.

Johnston, J. (1982). *Prevention in Human Services* 2 (In press)

Johnston, J. and Ettema, J. (1982). *Positive Images: Breaking Stereotypes with Children's Television*. Sage Publications, Beverly Hills, California

Johnston, J., Ettema, J. and Davidson, T. (1980). *An Evaluation of* Freestyle. Institute for Social Research, University of Michigan, Ann Arbor

Laosa, L. (1974). *California Journal of Education Research* 25, 303–309

LaRose, R. (1978). "Project *Freestyle*: Baseline Studies". Paper presented at the meeting of the International Communication Association

Lesser, G. S. (1974). *Children and Television: Lessons from* Sesame Street. Random House, New York

Levin, S. R. and Anderson, D. R. (1976). *Journal of Communication* 26(2), 126–135

Liebert, R. M. (1976). *Journal of Communication* 26(2), 165–171

Mays, L. (1977). *Formative Evaluation of New Pilot Segments for* Vegetable Soup II *and* Program 25, Vegetable Soup I. New York State Education Department

Mays, L., Henderson, E. H., Seidman, S. K. and Steiner, V. J. (1975). *On Meeting Real People: An Evaluation Report on* Vegetable Soup. New York State Education Department

Mielke, K. W. (1981). Personal communication

Mielke, K. W. and Chen, M. (1980). *Making Contact: Formative Research in Touch with Children*. CTW International Research Notes. Children's Television Workshop, New York

Mielke, K. W. and Swinehart, J. W. (1976). *Evaluation of the Feeling Good Television Series*. Children's Television Workshop, New York

Moody, K. (1980). *Growing Up on Television*. Times Books, New York

O'Bryan, K. G. and Silverman, H. (1973). "Research Report: Experimental Program Eye Movement Study". Children's Television Workshop, New York

Palmer, E. L., Chen, M. and Lesser, G. S. (1976). *J. Comm.* 26(2), 109–123

Paulson, F. L. (1974). *AV Communication Review* 22(3), 229–246

Pena, D. and Miller, G. (1971). "Social Skills Development in the Early Childhood Education Project". Technical Report, No. 7. Appalachia Educational Laboratory, Charleston, West Virginia

Reeves, B. R. (1970). *The First Year of* Sesame Street: *The Formative Research*. Children's Television Workshop, New York

Roberts, D., Herold, C., Hornby, M., King, S., Sterne, D., Whitely, L. and Silverman, T. (1974). "Earth's A Big Blue Marble: A Report of the Impact of a Children's Television Series on Children's Opinions". Institute for Communication Research, Stanford University, California

Rust, L. W. (1971a). *Attributes of* The Electric Company *Pilot Shows that Produced High and Low Visual Attention in 2nd and 3rd Graders*. Children's Television Workshop, New York

Rust, L. W. (1971b). The Electric Company *Distractor Data. The Influence of Context*. Children's Television Workshop, New York

Southwest Texas Public Broadcasting Council (1979). Caboodle *Project Evaluation*. Office of Education, US Department of Health, Education and Welfare, Washington

Sproull, N. L., Ward, E. F. and Ward, M. D. (1976). *Reading Behaviors of Young Children who Viewed* The Electric Company. Children's Television Workshop, New York

Strand, T. (1979). "Formative Evaluation of an ESAA-Funded Television Pilot Program on Desegregation". Paper presented at the meeting of the American Educational Research Association van Wart, G. (1974). Carrascolendas: *Evaluation of a Spanish/English Educational Television Series Within Region XIII*. Office of Education, US Department of Health, Education and Welfare, Washington

Williams, F. (1978a). "Project *Freestyle*: National Sites Results". Paper presented at the meeting of the International Communication Association

Williams, F. (1978b). "Sex-Roles on TV: More than Counting Buttons and Bows". Paper presented at the meeting of the American Psychological Association

Williams, F. and Natalicio, D. (1972). *Journal of Broadcasting* **16**, 299–309

Williams, F., van Wart, G. and Stanford, M. (1973). Carrascolendas: *National Evaluation of a Spanish/English Educational Television Series*. Office of Education, US Department of Health, Education and Welfare, Washington

Williams, F., LaRose, R. and Frost, F. (1981). *Children, Television, and Sex-Role Stereotyping*. Praeger, New York

Formative Research for *3-2-1 Contact:* Methods and Insights

Keith W. Mielke and Milton Chen

Introduction

This chapter addresses certain elements of the formative research for a television series called *3-2-1 Contact*, produced by the Children's Television Workshop (CTW). *3-2-1 Contact*, which premiered in January 1980 on the Public Broadcasting Service (PBS), dealt with motivating interests in science and technology among its target audience of children eight to 12 years of age. The first season of the series consisted of 65 half-hour programs, broadcast daily, Monday through Friday, over a 13-week period.

Each week focused on an interdisciplinary theme, such as Hot/Cold, Growth/Decay, and Fast/Slow, to present phenomena across a broad spectrum of scientific and engineering fields, tied together by their relevance to the weekly themes. Such themes, using everyday terms, were not stated in scientific or technological jargon, and spanned a wide variety of human experiences touched by science and technology. The themes offered a method of marshalling a considerable diversity of material available for treatment into a workable organisation, and provided a frame of reference for the various staff members who needed a common language in order to collaborate on the production. The themes brought together and focused more abstract scientific principles, and at the same time offered children multiple perspectives for looking at the world and themselves.

The series used a magazine format featuring three young people who were filmed both on location and in a studio setting. Animation and graphics were also utilised, as well as a mystery serial, The Bloodhound Gang, with three young detectives who solved mysteries using scientific thinking.

We will first discuss briefly a number of challenges posed in the course of planning and producing *3-2-1 Contact*, and how our in-house developmental, or formative, research responded to project needs through its various phases. Subsequent sections will then give an overview of a variety of research methods employed, some lessons we have learned about appropriateness of these methods for the target audience of 8–12-year-olds, and some general insights we have gained into their response to television and science.

The challenge to formative research

Formative research is typically contrasted with summative research or evaluation, as stated recently by Michael Scriven (1980), who in 1967 first coined the distinction:

Formative evaluation is conducted *during* the development or improvement of a program or product (or person, etc.). It is an evaluation which is conducted *for* the in-house staff of the program and normally remains in-house; but it may be *done by* an internal *or* an external evaluator or (preferably) a combination. The distinction between formative and summative has been well summed up in a sentence of Bob Stake's: "When the cook tastes the soup, that's formative; when the guest tastes the soup, that's summative". (p. 56)

Formative research for *3-2-1 Contact* was to incorporate the viewpoint of the target audience of 8–12-year-olds into every step of the production process, beginning with needs assessment and selection of program format, through production of individual segments and then completed programs. Both quantitative and qualitative methods were used, combining the efficiency of numerical data collected from large groups of children with the expressiveness of comments from individual children.

The needs of the in-house decision-makers, typically producers, required that the results be obtained as expediently as possible and communicated quickly and clearly. Timing, being keyed to production decisions, was critical. Those decisions would have to be made with or without the input of formative research, and it became the responsibility of the research staff to work within the discipline of the project schedule.

The more novel or difficult the production task, the more useful such formative research feedback can be to project management. As with CTW's other children's series that preceded this one – *Sesame Street* and *The Electric Company* – the idea of producing a daily children's series about science was both new and difficult.

For example, *3-2-1 Contact* was the first of CTW's productions to be designed from the very beginning for dual home and school use, requiring simultaneous concern for both the child viewing voluntarily at home and the teacher structuring a lesson in the classroom.

The organisation of the science content presented other fundamental issues. Science is everywhere. It can be seen in virtually every phenomenon and event. Criteria were required for the selection of science content. Policies and principles were needed for the structuring of this content for television treatment. Our aim was to organise the content coherently, but to avoid the formal and sequential nature of many classroom science curricula. The device of weekly themes was attractive for its ability to cross traditional scientific disciplines, to encompass a variety of scientific "big ideas" or principles, and to be exemplified through phenomena of interest to children.

On the receiving end of the programs, there stood a sophisticated and demanding audience of young television viewers, many of whom were showing declining interest in science. Didactic approaches and pedantic explanations would quickly alienate them. Production values would have to be competitive with those of popular commercial programing. In addition, the sheer magnitude of coordinating and producing 65 half-hour programs imposed its own requirements and constraints on production decisions.

In order to be responsive to overall project needs, the formative research agendas paralleled the activities of the project as a whole. The project evolved through four stages of development: needs assessment, format development, test show production, and series production.

Phases of research

NEEDS ASSESSMENT

During the earliest phase of needs assessment, formative research focused on questions such as:

What is the need for greater public understanding of science? How is this problem currently being addressed with children?

What is the status of science teaching in the elementary school?

What do surveys of science achievement indicate about children's knowledge of science?

What are their prevailing attitudes towards science and technology?

FORMAT DEVELOPMENT

During the next phase, researchers joined producers, writers, and science content experts in the crucial step of defining a format for the series. The research agenda expanded to include the conduct of actual field studies, evaluating a wide variety of existing and experimentally produced in-house material for questions such as:

What are the television viewing preferences of the target audience of 8–12's?

What attributes affect the appeal and comprehension of science presentations with children?

What formats would be most appropriate for the science series?

What types of characters do children find appealing?

What are children's perceptions of scientists and the work of science?

TEST SHOW PRODUCTION

The most critical and comprehensive sets of studies were devoted to evaluating the test programs which would serve as a model for the series itself. For this effort, we employed a "mosaic" approach, where multiple research designs were applied simultaneously. Data collected through one method were complemented by data from others so that a collective evaluative picture emerged. Four areas for investigations were:

What is the appeal of the various segments and cast members?

What points of information and overall concepts are more and less comprehensible to viewers.?

What is the competitive appeal of the series in a home-viewing environment?

What is the series' potential for classroom utilisation?

SERIES PRODUCTION

During series production, formative researchers functioned as representatives of the target audience in daily consultations with Production and Content staff, drawing on what was by that time an extensive accumulated knowledge of the reactions of the target audience. Selected programs from the first four weeks of the series were also evaluated in the field as they were completed.

In all, the formative research program comprised more than 50 field studies with over 10 000 target-age children. The following highlights draw upon that experience in summarising some of what we have learned about formative research methodology and our target audience. While the original purpose of the formative research was to guide Production, and was consequently rather stimulus-specific to the materials being evaluated, the overview here, directed towards both methodological and substantive insights, will deal with issues of more general applicability.

Formative research methods used with *3-2-1 Contact*

The formative research methodologies surveyed here are organised by the type of questions they addressed and the functions they served. They range from techniques borrowed from previous research to original designs. They have in common the fact that each was adopted and/or adapted for use on this project according to its responsiveness to specific research questions, feasibility for use with children in the target audience, and capacity to provide quick feedback to producers. Typically, several methods would be combined within a single research setting.

Many of the methods described have potential for further combination, extension, and modification in other research settings. Single tasks can frequently be addressed with different methodologies, and single methodologies can frequently be adapted to serve different tasks. The uses of these research methods with *3-2-1 Contact*, therefore, are considered neither exhaustive nor prescriptive.

CHILDREN AND SCIENCE

In early stages of the project, some of the questions asked by the Production staff dealt with the status of the target audience and their relation to science. What do they know about science? What vocabulary do they use to express questions they have? What imagery do they have of science and scientists? What topics are they most interested in? Such questions were addressed with the following methodologies:

Photograph study
Open-ended question generation
Hands-on exercises
Content analysis of science essays

The four methods are described below.

Photograph study

This method responded to the need of the project team to obtain a measure of children's interests across a wide variety of science topics. Presenting children with photographs and a one-sentence descriptive caption seemed the most "neutral" way of presenting a topic. Based on our experience with testing television material, we found that it became analytically difficult to differentiate between viewers' response to production treatments versus appeal based "purely" on the topic itself.

In this method, a series of photos (e.g. 30) collected from a variety of magazines was presented to large groups of children (e.g. from 20 students to an entire classroom). Each photo showed a science topic (e.g. supertanker, whales, dinosaurs, space colony). Each child was given a questionnaire listing the titles of the photos, organised into groups of six. The photo was shown to the children and a one-sentence caption, factual and evaluatively neutral, was read aloud. Asked whether they would like to know more about the topic (read a magazine story about it), children marked the appeal of the topic measured on a four-point scale from "definitely yes" to "definitely no". At the end of every sixth photo, children were asked to review the entire group of six and circle the one they would most like to read about and cross out the one they would least like to read about.

Two types of data were then analysed: (1) average appeal score per photo, for total sample, age, and sex; (2) percent of "most" and "least like to read" votes from groups of six, for total sample, age, and sex.

Open-ended question generation

A frequent request from producers during the project was "What science topics are kids interested in?" If such universally high-appeal or low-appeal topics existed, the series could take advantage of preexisting appeal in incorporating high-appeal topics into the series.

It became clear in the first few studies that production treatment of a topic, rather than the topic itself, was a more powerful predictor of appeal. Two different treatments of the same topic (e.g. predator–prey relationships) might test well in one instance and poorly in another. These findings helped to qualify the interpretation of appeal data. High appeal topics could not, in any determinable way, be incorporated into the program without serious consideration of appropriate treatment. Likewise, low appeal topics should not be rejected, since much depended on the way in which the topic was posed.

The *Photograph study* was one method for offering children as neutral a stimulus as possible to elicit their reactions to the appeal of topics. A more lengthy television treatment (of even 30 to 60 seconds) contains many more cues and possible sources of idiosyncratic appeal (or lack of appeal) that apply more to treatment and not to the topic itself. These types of issues led research to question whether the notion of topic appeal independent of production treatment was a useful one.

A different purpose evolved in our attempts to study the topics of children's curiosities. A more direct production purpose became evident in the need to know at what level the cast should pose their questions on the show. Also a specific department within the magazine format devoted to "Kids' Questions" was considered for production.

For these purposes, some information on questions children found interesting and the particular phrasing of such questions was needed. For instance, we were interested in whether children were attracted to rather simple and straightforward questions found in science textbooks ("Why do seasons change?") or to more fanciful questions ("Why does water come out of the faucet in a drop instead of some other shape?").

Tapping the honest and natural curiosities of children is not simple from a methodological standpoint. Many alternative ways of designing a study were discussed. We considered whether responses collected in a school setting for such a study would take on a school-appropriate or teacher-approved nature. On the other hand, conducting the research in homes or after-school settings presented other problems of logistics, parental approval, and diminished sample size. We also addressed the issue of whether "outsiders" like ourselves could obtain candid reactions to the curiosities of children and whether some more unobtrusive measure such as parents' recording of questions asked by their children might prove more successful.

A final problem involved the placement of children in an appropriate "question-asking" frame of mind. Even adults might be challenged to name three specific questions they have wondered about or would like to know the answer to. We decided to employ a design that would allow children to react to *models* of such questions and then generate questions of their own personal interest.

In this design, samples of children were given a questionnaire listing 20 questions in groups of four. They were asked to indicate the question in each group of four they "most wanted the answer to" and the one they "didn't care about at all". They were then asked to write on the back of the sheet their own questions about the world – anything they had ever wondered about or would like to know more about. A subsample was interviewed at greater length for the reasons behind their responses.

Hands-on exercises

At one point in the project, the concept of surface-to-volume ratio was selected for possible treatment in the Big/Little week. While the concept of volume increasing at a higher exponential rate than surface seemed to be a difficult one for the audience of 8–12's, Content scientists felt that the idea was of such scientific significance that introducing it to children in the series might lay important groundwork for later development of scientific thinking skills.

Specific existing materials could not be located that resembled Production's ideas for treating the concept. Working with the Content scientists, Research developed a series of six simple hands-on manipulations to observe children's ability to articulate the surfaces and volumes of objects and to make connections between the two. One exercise used modelling clay in a manner similar to Piaget's conservation tasks and asked children to decide whether changes in the form of an object resulted in changes in the amount of surface or volume. Other exercises involved estimating the number of sugar cubes required to double the outside dimensions of a single cube. The responses of about 20 children were analysed for particular difficulties, both conceptual and semantic, with the concept of surface-to-volume ratio.

Content analysis of science essays

One of the basic questions to be addressed in the formative research involved the existing attitudes of children toward science and the work of scientists. Information on questions in this general area had important implications for guiding the presentation of scientists on the series as well as for maximising the appeal of specific content. A study was designed to gather such information directly and in an unobtrusive manner.

Teachers in a school in upstate New York assigned their 4th, 5th and 6th grades a topic for a composition: "Why I Would (Or Would Not) Like To Be A Scientist" or "A Typical Day In The Life Of A Scientist". Students were free to choose either topic. Once the teachers had graded the compositions for their own classroom purposes, such as spelling and grammar, they were sent to CTW for content analysis.

Content analysis is a method of obtaining an objective characterisation of a body of content within the constraints of the decision rules employed. In this analysis, content was coded according to references made to individual science categories such as archaeology, botany, biology, or chemistry. References were then classified according to whether they were positive, neutral, or negative. Tables were constructed to display the data according to frequency of assertions, by total sample, age, and sex. The analysis acknowledged that the content of the children's essays would reflect to some degree their current classroom activities. Data from this content analysis presented a picture of children's attitudes consistent with more anecdotal evidence gathered continuously in the field.

GENERAL TELEVISION VIEWING HABITS AND PREFERENCES

This category of research methodologies was used to address questions of television viewing habits and preferences at a general level, obtaining reactions to verbal references to programing rather than reactions to actual program material. Some of the *3-2-1 Contact* questions here included the following: How popular are game shows among the target audience? How familiar is the target audience with existing science programing, and how does it compare in popularity with entertainment fare commercially available on television? How do entertainment programs using a magazine format compare in appeal with entertainment programs using a plotted drama format? For such questions, the following methodologies were employed:

Television interest survey
Ratings analyses
Television guide excerpts

These are described below.

Television interest survey

A basic consideration in the early stages of developing the series was to obtain information on the types of programing viewed by the audience. Our interest was in the kinds of programs, formats, and characters that were popular with children in our audience. We were especially interested in different patterns of interest by sex, age, or ethnicity. This information had implications for the important decisions of format selection, as well as casting, scheduling, and production approach to individual segments.

To obtain more detailed breakdowns of viewer preferences by age, sex, and ethnicity, tailored to project-specific questions, the Research staff found it necessary to design a national survey of children's viewing preferences (Mielke, Chen, Clarke and Katz, 1978).

An eight-page questionnaire was designed containing items on: (1) one favorite show; (2) shows viewed yesterday; and (3) awareness and viewing patterns for 20 programs, selected because of interest in their content or format. Through graphics and layout, the questionnaire was designed to be fun to complete, and to not look like a test. Reproductions of photographs from the 20 programs were seen next to each item to cue memory. Line drawings symbolising response options also minimised the reading ability required.

The *Television interest survey* questionnaire was distributed to 4148 children around the US. Simple statistics such as frequency distributions and rank orders were computed for the items.

Ratings analyses

Occasionally during the needs assessment and format development phases, producers desired information on the ratings of various shows. For example, there was at one time interest in the game show format, and Nielsen ratings for the viewership for various game shows among 6–11-year-olds (the closest Nielsen category to our target audience) were analysed by the Research staff.

Television guide excerpts

Like the *Television interest survey*, this method was specifically tailored to the needs of this project. The method was first developed in response to Production interest in the relative appeal of magazine-formatted shows *versus* plotted dramatic shows. Specifically, there was interest in how four particular magazine/variety shows fared against their plotted drama competition.

The actual timeslots for the four programs of interest were reproduced from a recent television guide. All programs competing in that timeslot were shown, as they were in the *TV Guide*. Children were asked to circle the one program they would "usually watch" on that day at that time. If they did not usually watch television during that time, they were instructed not to mark any show.

Because of the ease of administration of this questionnaire, responses from large samples of children could be obtained (i.e. over 250 children per study). Many children indicated that they often used the *TV Guide* to make viewing decisions. Respondents for two such studies, while demographically different, indicated similar viewing patterns for the four time periods.

MEASURES OF APPEAL ACROSS PROGRAMS

A substantial portion of the formative research effort was devoted to various aspects of assessing appeal and comprehension. Before *3-2-1 Contact* materials were produced, appeal and comprehension testing was conducted on a variety of existing materials produced externally. Later in the process, the focus shifted to segments and completed shows produced in-house.

The unit of analysis on appeal questions varied across a range of entire programs, segments within programs, and portions within segments. Methods that engaged the broader focus – measures of appeal across programs – are described in this section, and include the following:

Program comparisons
Triplet voting
Meta-research on documentary films

Program comparisons

During the course of the project, this method was applied to comparisons between two (paired comparison) and four (quartet comparison) programs. Using a forced-choice format, viewers were asked to view the pair or the quartet and select the one they found most appealing. In the case of quartet voting, children were also asked to select the least appealing program.

One paired-comparison design involved three pairs of three magazine programs. Each was paired with the other two, and viewers saw approximately twenty minutes of each program. The quartet design was used in two studies; one compared the appeal of four Bloodhound Gang mysteries, the other compared one episode of the Bloodhound Gang with three other programs. In each *Program comparison*, the length of the programs being compared was edited to be approximately equal and total viewing time was kept to a maximum of 40 minutes to allow for administration of the items within a one-hour period.

Triplet voting

This technique grew out of the need to design a study that would gather data on both the appeal of individual segments and the appeal of topics. It has the further function of modelling the program-selection behavior of children in choosing programs for viewing based on a first impression of the material seen at the moment on the screen.

Nine one-minute excerpts from science shows were organised into three triplets, each showing an example of a space, animal, and micro-organism topic. After viewing each triplet, children selected the most- and least-liked segment in private "Voting Books". After viewing all nine excerpts (three triplets) they again made most- and least-liked choices from among the entire group of segments.

Meta-research on documentary films

Given the number and range of formative research studies conducted for this project, the Research staff was periodically asked to summarise research results for presentation to staff. One specific request asked the researchers to look across the types of documentaries evaluated and arrive at some conclusions as to their relative appeal, and, especially, reasons for their appeal.

This request suggested a form of meta-research in which the researchers synthesised the results of several research studies instead of concentrating on the results of testing any one individual program. Twenty-three documentary films had been evaluated at the time of the request. Using their knowledge of the quantitative and qualitative data on each film, three researchers independently grouped them into seven categories from most to least effective. Differences in ranking were resolved by discussion until consensus was reached.

Beyond those rankings, the more useful analysis involved researchers' speculations on the attributes explaining the response of children to the appeal and comprehensibility of documentary films. Brief evaluations of strengths and weaknesses of each of the 23 films were written, as well as more general discussion of guidelines and recommendations to producers and documentary filmmakers.

MEASURES OF APPEAL WITHIN PROGRAMS

Several methodologies in this category were used, examining relative appeal of segments within programs and even of scenes within segments. The following methods will be discussed:

Program analyser
Segment voting
Small-group interviews
QUBE cable television study
Cast appeal studies

Cast appeal studies are difficult to place on any macro/micro continuum, because test audiences make a summary judgment of comparative cast appeal based on whatever program material they have been exposed to: multiple shows, single shows, or single segments. With the exception of *Cast appeal* studies, all of these methods were used to assess relative appeal of segments within programs. In addition, the *Program analyser* and *Small-group interviews* can be fine-tuned to assess appeal of material within segments. The *Program analyser* is capable of generating quantified appeal data on a moment-by-moment basis in units of time down to a fraction of a second, although we did not find such precision to be necessary. As such, it permits segment comparisons as an aggregation across micro-level units of analysis. The level of specificity is, of course, shaped by the questions asked.

Program analyser

The *Program analyser* technique, developed by Frank Stanton and Paul Lazarsfeld, is used in various forms for audience testing of television programs. Test audiences give evaluative reactions by such means as button-pushing, as a program is being viewed. Since the target audience for *3-2-1 Contact* was older than for *Sesame Street* or *The Electric Company*, some version of *Program analyser* methodology appeared relevant to our research interests. Would 8–12's understand the instructions and respond reliably? If so, the technique promised to be of major utility in evaluating the appeal of a wide range of programing. The Research staff devoted substantial effort to refining the technique for application with 8–12-year-olds. In so doing, a new generation of *Program analyser* was designed. A computerised *Program analyser* and an earlier wired version were used in the formative research, and both are described below.

Wired program analyser. CTW's initial *Program analyser* system involved twenty buttons attached through a wiring harness to an event recorder. Twenty heat-sensitive styluses recorded periods of button pressing on a moving paper scroll in the event recorder. The scroll was coded manually to generate a graph of program appeal over time.

The typical *Program analyser* instruction to adult respondents asks each individual to evaluate both positive and negative aspects of the program. With children, this instruction carried two methodological problems: (1) difficulty in keeping dual dimensions simultaneously in mind; (2) social desirability bias towards giving more favorable evaluations, especially in a school setting.

Based on experimentation with different voting instructions with our target audience of 8–12's, the most useful approach was to divide each test group of 20 students into two halves. Ten children were given green buttons and told to press for portions of the show that were "interesting or fun to watch". The other ten were given red buttons and told to press the buttons when the show was "boring or not fun to watch". These instructions (1) reduced the complexity of the task for the child; (2) legitimised the negative response; and (3) provided a built-in validity check. Interpretation of the two graphs, plotted every 12 seconds, was based on cross-validation between the two opposing groups. As "interesting" votes rise, "boring" votes should logically fall, and vice versa. Repeated verification of this phenomenon across test material was strong evidence that the instructions were tapping a common and reliable indicator of program appeal.

The end product of this system was quite useful in pinpointing moments of high and low appeal within a program. However, the wiring harness was cumbersome to use in the field and coding of the scroll was time-consuming, taking a team of two people about fifteen hours each to transform the scroll into an interpretable set of graphs for the total sample, boys/girls, and younger/older children.

Program Evaluation Analysis Computer (PEAC). Work began in the summer of 1978 to design a portable, wireless *Program analyser* with completely self-contained viewer response units. It appeared at that time that one of the new generation of personal computers could substantially decrease the time for data analysis. Development of the system moved forward in partnership with the Ontario Educational Communications Authority (OECA) and a computer design engineer. The final system was delivered in January 1979 and was used extensively for evaluation of test shows on which the *3-2-1 Contact* series would be modelled.

The key components of the *Program Evaluation Analysis Computer* (PEAC) are:

1. A 48K Apple II microcomputer with two disk drives;
2. 40 individual hand-held units, containing a microprocessor and timing crystal for logging in responses;
3. A color television monitor for on-screen color display of data;
4. A dot-matrix printer for hard-copy printout.

With the PEAC system, each child in a viewing group as large as 40, but more typically a classroom of about 25, was given one of the wireless, battery-powered hand units and told to respond according to instructions previously described.

The hand units contain a 16-key keyboard for responses. During the program, half the viewers in each group pressed key "A" whenever they felt the program was "interesting or fun to watch". The remaining half pressed key "D" for those portions that were "boring or not fun to watch".

Viewer reaction was sampled by the hand units every 10 seconds using a "majority vote" decision rule, i.e. a key pressed for 5 seconds or more out of a 10-second interval was logged in the memory of the hand unit as the response of record for that particular interval. A key pressed for less than 5 seconds was logged as no response.

Following viewing, students typically responded on the units to a series of multiple-choice questions regarding their favorite segments, cast members, or comprehension of program material. At the end of the test material, the hand units, having been programed to do so, automatically switched from a timing mode to a question mode during which responses to the questions were logged.

Typically, questions were presented to each viewing group on large poster boards at the front of the classroom. Large black-and-white photographs of the cast members were also presented in conjunction with the cast appeal items.

At the end of each test session, the hand units were collected and the data transferred to an Apple II microcomputer and a magnetic diskette for analysis and storage.

The PEAC system has four major advantages over the previous *Program analyser*:

1. Convenience of a wireless hand unit for large-group testing.
2. The ability of the units to log responses to multiple choice questions, removing the barrier of responding by paper and pencil. (This mode operates in addition to the timed mode used for evaluative reactions.)
3. Immediate feedback of results, made possible by the computing power of the Apple computer. The period from data collection to final response graph was cut from thirty hours to fifteen minutes.
4. The ability of the Apple to generate data presentation formats such as profiles or histograms on a color television monitor. Such displays can be viewed in time to the actual test program. Producers can examine the response to each 10-second interval on one television screen with the

corresponding stimulus material played simultaneously on another. This feature of the PEAC system opens new opportunities for communicating research results in a language and format attractive to television producers and writers. The system made it possible for Research to both collect a wealth of useful data across a number of important questions and to meet Production's demand for quick turn-around time in reporting of results.

It should be mentioned that the above use of the PEAC system is only one of a range of applications possible. For instance, the sampling interval for collecting data can range from 0.25 seconds to as long as several minutes. Decision rules for logging the data can use either the "majority vote" rule or the criterion of the last key pressed in the interval. Response instructions and key-pressing can be varied to suit the research task. The range of PEAC applications has yet to be explored fully and holds promise for improving the speed, reliability, and convenience of both television and print research.

Segment voting

The *Program analyser's* moment-by-moment data can be supplemented via *Postviewing segment voting*. Each child is given a sheet listing the segments viewed and asked to choose the best- and least-liked segment. Such data give a final index of relative appeal within a show, based on a viewer's retrospective look at the entire program. *Segment voting* can be especially useful in discriminating between segments that received similarly high and low *Program analyser* responses while viewing.

Small-group interviews

Large-group studies, such as the *Program analyser*, typically used 100 or more children. There was also a need for complementary data: broader, less structured, and more in-depth. The difficulty of obtaining such data meant that small samples of children would be used. The *Small-group interview* of one researcher working from a prepared questionnaire with three or four children was our main methodology for obtaining this more impressionistic but equally valuable type of data. It also became the method of choice for gathering most of our information on comprehension since, unlike paper-and-pencil items, it is not dependent on children's reading ability, nor is it entirely prestructured.

Instead of generating quantitative data, such as percentages of correct responses for points of comprehension, we were more often concerned with underlying reasons *why* children were not understanding a particular point or *why* they liked a particular character. This type of information was most easily obtained through face-to-face interviews where a trained researcher could probe for children's underlying perceptions, attitudes, and understanding. We found that the 5th- and 6th-graders were especially responsive and articulate in these situations.

Using this method, researchers recorded the *verbatim* responses of each child in the group on the prepared questionnaire. These responses were later submitted to an informal content analysis, with majority and minority viewpoints noted and represented in the Research memo. Individual comments, representative of a larger pattern of response, were reported *verbatim* to Production and Content.

Typically, this type of data, "straight from the kids", is more easily interpreted by producers and other nonresearchers than quantitative types of data. While child comments attract the attention of producers, they also require vigilance from the researcher in assigning proper value or importance to particular comments.

QUBE cable television study

While the *Program analyser* and *Small-group interview* methods attempt to simulate a natural viewing setting, they nevertheless operate most typically within the constraints of the school classroom. The QUBE Interactive Cable System in Columbus, Ohio offered a unique opportunity to study viewer response under actual home viewing conditions.

The QUBE System is an interactive cable system enabling viewers in their homes to respond to programing. In addition to the buttons for program selection, the cable console in each home also contains five response buttons. Questions with up to five response options may be posed to the home-viewing audience, with a master computer tabulating these responses in a matter of seconds throughout the entire system of 30 000 subscribers. It is possible to restrict the viewing sample for a test program by sending the cablecast to preselected households. This capability is called a "narrowcast" and was employed in this study.

The most important aim of this study was to find out the age and sex distribution of the sample that, once informed of the cablecast of *3-2-1 Contact* test shows, voluntarily tuned in. Data were also collected on specific appeal and comprehensibility questions to provide cross-validation of similar data collected in other studies.

Through random selection from a listing of QUBE subscriber names, a total of 737 children in the 5–15-year-old age range was invited to participate in the actual study. One child and his/her parent in each household were informed of the narrowcast. Each child was also invited to view the program and participate in the questions following the program.

Two test shows were selected for this study, and were narrowcast to the 737 QUBE households on consecutive days. The narrowcasts lasted approximately one hour including lead time, the half-hour program, and postviewing questions.

Both before and after the show, age and sex identification questions were asked of the audience by a live host at QUBE. After viewing, segment appeal and character appeal questions were asked using a 4-point scale from "very interesting" to "very boring".

The response options appeared on the screen as the host read them, accompanied by the appropriate key number to be pressed on the home console. A freeze-frame from each segment appeared on the screen as the host directed viewers to respond to that segment. Several seconds of response time were provided for each question, during which the phrase "Touch Now" flashed on the screen. This "Touch Now" technique was used throughout the testing.

Segment appeal testing was followed by several multiple-choice questions focusing on the program content. Each question and response option appeared on the screen. This set of questions was followed by studio cast and Bloodhound Gang cast appeal questions.

The final set of questions involved home-viewing patterns. A series of multiple-choice questions was asked, to determine whether the child was viewing with other people and whether those people were parents, siblings, and/or friends. Other questions asked the audience whether they would watch again if given the opportunity and what audience age range they perceived as appropriate for the show.

During the narrowcast, the QUBE computer recorded data from the viewing audience. In addition to registering the raw number of viewers at various intervals during the program (approximately 20 seconds apart), the system also provided response rates for each question in the testing and cross-tabulations of all questions by respondent age and sex.

Cast appeal studies

During the test show evaluation, when the appeal of the cast members was a basic question, formative research data were collected through several methods. One method operated as part of the larger *Program analyser* studies using the PEAC (Program Evaluation Analysis Computer) system. Large black-and-white photos were shown to children at the end of the program. Each cast member was assigned a number and children logged responses for their most- and least-favorite cast member on their hand units. Their votes were then tabulated by the Apple computer for individual programs and subject age and sex.

During *Small-group interviews*, children were handed a sheet showing small black-and-white photos of the cast and asked to indicate their favorite and least favorite cast member. These data were entered into the computer and merged with the data from *Program analyser* samples. Children in the *Small-group interviews* were then interviewed in depth for their perceptions of the cast.

MEASURES OF COMPREHENSION

Measures of comprehension address such questions as: What is perceived? What is remembered? What terminology is and is not understood? What processes of information acquisition are going on as the program is being viewed? What information gain is demonstrable in a postviewing test? These types of questions were addressed with these methodologies:

Small-group interviews
Program analyser
Freeze-frame comprehension testing
Fill-in-the-track comprehension testing
Posttest of viewers' comprehension

The use of *Small-group interviews* and the *Program analyser* operating in "multiple-choice question" mode have been discussed in previous sections. They apply as well to the measurement of comprehension, but will not be discussed further here. The other methodologies are described below.

Freeze-frame comprehension testing

This comprehension-testing technique borrows from a method first used with formative research for *Sesame Street* and *The Electric Company*. Using the freeze-frame capability of the videotape player, this method tracks a child's comprehension of a program or segment. During certain moments or events in the program, the researcher may be interested in the child's understanding at that particular time. It therefore is appropriate to query the child for his or her understanding at those particular points in time.

With three or four children per viewing group, the segment was played up to the point or points in question, and the videotape was stopped with the relevant frame frozen on the screen. The researcher then asked the children for their understanding of the segment up to that point, using the frame as a referent for discussion. The method can be adapted to research on films by simply turning off the projector, sacrificing the freeze-frame capability of videotape. Analysis of viewers' comments indicated points of understanding and confusion and helped to suggest strategies for improving presentation of information.

Fill-in-the-track comprehension testing

Like *Freeze-frame testing*, this technique was also borrowed from previous CTW research. Its rationale is also based on an attempt to track children's comprehension of program content in a segment.

Viewers were first shown the segment in its entirety, with children informed they would subsequently be asked to perform as the film's narrator during a second showing. During the second time, with the sound turned off, individual children's narrations "filling in the audio track" were recorded by hand as well as on audio tape for later analysis. The comparisons of their narrations with the *verbatim* transcript of the film provided interesting contrasts showing areas of understanding and misunderstanding as well as difficult vocabulary for the target audience.

Posttest of viewers' comprehension

At one point in the test show evaluations, it became necessary to take advantage of the same set of respondents for dual purposes, due to time constraints and access to subjects. One purpose was to obtain information on comprehension of show messages, on a program-by-program basis. The second purpose was to develop an exploratory measure of information gain that could be attributable to viewing the week of five programs, seen Monday through Friday. From experimental reasoning and prior experience, we believed that a bias would be introduced if the same subjects who answered daily comprehension questions also provided the sample for measurement after the Friday show of overall information gain from the set of five shows. The daily questions might cue viewers to look for certain types of information in subsequent shows, guessing they might be responding to subsequent comprehension items. The daily questioning was a form of "rehearsal" that a more casual viewer might not employ. Even pretest sensitisation without daily follow-up questions might bias the results. This potential contamination made the group tested for daily comprehension unsuitable for participation in an experimental test of information gain.

The standard posttest-only design, applied to this situation, would have followed this procedure:

	Monday through Friday	Friday
(Group A) Nonviewing sample	—	Posttest
(Group B) Viewing sample	View 5 shows	Posttest

Within the limits of statistical error, differences in posttest scores between Group A and B would be attributable to the viewing experience, if assignment to the groups was random. This design, however, did not make the most efficient use of the Group A respondents required by the time constraints during the test show evaluation period.

Our solution followed the logic but not the exact procedures of a "posttest only" design for the information gain study. The second study – daily comprehension measures – was superimposed on the first study. The two studies that were conducted simultaneously can be separated conceptually. The information gain study followed this design logic:

	Monday	Monday through Friday	Friday
Nonviewing sample (Group A)	"Posttest"	—	—
Viewing sample (Group B)	—	View 5 shows	Posttest

The daily comprehension study followed this design logic:

	Monday through Friday
Group A	View 5 shows, responding to daily comprehension items

When combined, the design of the study is as follows. Within several classrooms, respondents were assigned randomly into Group A or Group B.

	Monday	Monday through Friday	Friday
Nonviewing sample (Group A) N = 100	"Posttest"	View 5 shows responding to daily comprehension items for different study	—
Viewing sample (Group B) N = 103	—	View 5 shows: no intervention	Posttest

The "posttest" for Group A was not functioning as a pretest, because there were no pretest-posttest comparisons. For the purposes of the daily comprehension study, Group A then became the sample for viewing the five programs and responding to daily comprehension questions. Group B received no tests before or during the sequence of viewing, but was tested after viewing all programs. The "double duty" payoff came in comparing the scores for Group A with Group B in a study of information gain *and* in analysing Group A's response to the daily comprehension questions. The elapsed period of four days between Group A's "posttest" and Group B's posttest was brief enough so that history or maturation were not considered threats to internal validity.

Methodological issues

The research methods described previously run the gamut from small samples to large samples, from quantitative to qualitative, from formal to informal, from old "tried and true" methods to more novel approaches. The methods were applied in this case to children 8–12 years of age, the medium of television, and the subject area of science. From this experience, we will set out below some of the methodological insights gained.

PAPER/PENCIL MEASUREMENT

One advantage of the 8–12 target audience was their ability to make more use of questionnaires and other forms of paper/pencil measurement. Nevertheless, great care was needed in the design of such instruments because children of 8–12-years-old are not sophisticated in the conventions of questionnaires, and levels of verbal fluency and literacy vary greatly.

One of the most successful instruments was designed for the *Television Interest Survey*. The use of photographs to identify televison programs, in addition to the verbal reference, was helpful. The restriction of the number of response options to three aided simplicity, and we feel that the reinforcement of the meanings of the response options with stick-figure drawings enhanced the validity of the responses. As an internal check on the proportion of non-meaningful responses, we included a non-existent television show, to which the same three response options applied:

"I have never heard of this show."
"I know about this show, but don't usually watch it."
"I watch this show whenever I can."

Fortunately, the reported viewership of this non-existent show was very low, ranging from 2% of the sixth-graders up to 4% of the third-graders.

POSITIVE RESPONSE BIAS

Children in this age range have discovered the social value of being able to infer behavioral expectations from parents, teachers, and other adults. This social skill frequently presents a problem for the television researcher by taking the form of a "positive response bias", the hesitancy to say anything critical of the test materials in order not to appear discourteous.

We built in safeguards against such bias through research designs that legitimised the negative response and through constant awareness in the interpretation of results. For instance, our research teams were careful not to state that they represented CTW in working with children. To our surprise, many children were able to identify CTW's name and its programs and held favorable attitudes towards the organisation. We stressed to children that we were not actually the creators of the material, so that they could feel free to offer criticism without concern for offending anyone present. We also went to considerable lengths to explain that candid and thoughtful reactions, positive or negative, were of the most value to us.

FORCED-CHOICE *VERSUS* SCALE ITEMS

Awareness of the potential for positive response bias led to other preferences in methodology as well. For instance, we often evaluated sets of television segments or cast members. If a graduated scale were applied (e.g. a 5-point scale ranging from "like a lot" to "dislike a lot"), the respondent would be free to give a high or favorable rating to each segment or character. Such behavior not only avoids the social risk of being negative, but also permits one to escape the responsibility of making judgmental discriminations and applying meaningful criteria.

Because of this tendency on the part of respondents, we generally preferred so-called "forced-choice" questions such as: "From this set of three TV segments (or cast members), circle the one you like best. Put an X through the one you liked least". Such procedures elicit a process of discrimination from the viewer, and legitimise the negative response by specifically asking for it.

EXPERIENCES WITH THE PROGRAM ANALYSER

In giving *Program analyser* instructions to children, we have learned to avoid the words "like" and "dislike", because children perceive dual meanings for those terms. On the one hand, a "dislike" vote might mean that the material was boring and not engaging their interest. On the other hand, the test material could be quite interesting or compelling, but "disliked" because viewers did not approve of the actions taken by the individual characters.

In using the *Program analyser* across a wide variety of materials we have found the technique generally to be more successful at tracking the relative appeal of segments with a magazine format than across longer plotted dramas. Appeal profiles for dramatic presentations often built to a point where no further discriminations or changes in appeal were evident.

We hypothesised that, unlike viewing of a magazine format, the viewer of a plotted drama makes a different type of psychological "investment". Once the threshold of commitment has been reached to stay with the program to its resolution, the viewer finds it more difficult to make moment-by-moment judgments of scenes whose full interpretation rests on the final denouement. With the shorter segments of a magazine format, judgments are more independent and instantaneous. Appeal profiles often changed dramatically with transitions to different segments.

Selected findings and insights

Along with Production and Content, Research was part of a three-unit team with major responsibility for series development, and most of the research output was of a project-specific nature, applying to decision-making needs as they arose in the course of the project. Some generalisations, potentially of broader interest, will be summarised here. Additional information can be found in Mielke and Chen (1980, 1981), Chen (1980–81), and Myerson Katz (1980).

GENERALISING FROM FINDINGS

Our formative research experience has increased our sensitivity to stating generalisations from findings without appropriate cautions. Television production can have enormous variations in the method of presentation for any single topic. The interaction of topic appeal and method of presentation thus makes it difficult to generalise.

Consider, for example, the question of whether snakes, as a general topic, are appealing to children. On one occasion, a short segment of a snake swallowing its prey was repulsive to children. On another occasion, similar footage, set in the context of predator–prey relationships surrounding a termite mound, was highly compelling, eliciting much discussion both during and after viewing.

Acknowledging the difficulty of stating generalisations that will apply across all cases, we attempt here to offer a series of insights that have emerged most frequently from our fifty studies and our sustained contact with the target-age group.

GENERAL CHARACTERISTICS OF 8–12's

Television interests and preferences

Children in this age range are familiar with a broad range of television programs, formats, and characters, spanning situation comedy, action/adventure, game shows, feature films, and commercials. They hold high standards for production values, often stating the need for "action" and "good acting" and criticising material that is "corny" or "silly". The shorter attention span that characterises younger children is less of a concern with this audience, many of whom regularly watch movies.

Given the range of programing viewed by this audience, they are in surprising agreement on their favorite shows. Our *Television Interest Survey* indicated that only seven programs, such as *Charlie's Angels*, *Happy Days*, and *The Incredible Hulk*, accounted for more than half of the unrestricted "favorite show" nominations from more than 4000 children.

Scientific thinking and knowledge

Children of this age tend to think in concrete terms and give literal explanations based on their own experiences with the visible physical world. They do not readily make abstractions. They are less familiar with the unseen world, such as molecules and micro-organisms, as well as invisible larger processes, such as water and energy cycles. Therefore, in *3-2-1 Contact*, careful attention was devoted to vocabulary and explanations. Prior knowledge on subjects that adults take for granted, such as the gas crisis or computers, cannot simply be assumed with children.

However, many children do have pockets of knowledge based on home, school, and viewing experiences. They often have more grounding in the biological world of animals and plants than in the physical world of natural phenomena. Animal behavior and the human body are areas of high interest and curiosity.

Sex differences

Stereotypical sex differences were found using multiple methodologies, both confirming the need for the series and indicating the difficulty of expecting to alter such interests and attitudes on a large scale.

For instance, in our *Television Interest Survey*, girls were more attracted than boys to shows characterised by themes of warm human relationships, often in family situations, featuring women in leading roles. Boys showed a relative preference for action/adventure programs with male lead characters. In the *Triplet voting study*, girls preferred brief excerpts of animal films while boys preferred those with an outer-space setting.

FACTORS AFFECTING APPEAL AND COMPREHENSIBILITY

Format preferences

Early studies demonstrated a clear preference for plotted drama over the segmented magazine format as a general program type. The format where a problem is posed and resolved through relations between various recurring characters is a powerful device for attracting and holding viewers of this age. This trend continued in children's reactions to The Bloodhound Gang, which used this finding to advantage by posing mysteries solved by young detectives slightly older than the target audience. These mini-mysteries consistently tested as the highest appeal element of the series. The challenge to researchers and producers is to investigate new ways of incorporating instructional content into the plotted drama format.

Documentary storyline

The appeal of plotted drama can be translated into documentary film segments as well. Often the dramatic development of a problem and its resolution can help motivate the need for a scientific approach or piece of information. Children were particularly drawn to those documentary segments with emotional life/death themes, for example, stories of endangered species or premature babies.

Active visuals

Children of this age appear to process information from television predominantly from the visual channel and secondarily from the audio. Unusual or action-filled pictures of phenomena were of high interest, such as those of the world's largest pizza or a massive oil spill. Materials with static visuals, which relied heavily on the audio track to carry the informational load, were unappealing.

Explicit connections between segments

Contrary to some initial hypotheses, children did not view as didactic a clear statement of educational "headline" tying together each day's segments. In fact, when such connective material was not present, viewers did not perceive any unifying theme and viewed the magazine format as an assortment of "just a lot of different things".

Appropriate humor

Eight- to twelve-year-olds possess an appetite for humor. They look for opportunities to laugh while viewing, especially during animation, and are disappointed when their expectations are not fulfilled. They were quick, however, to label attempts at humor that were "silly" or "babyish".

Competence

In their reactions to characters, whether scientists or young people, children consistently favored those who "did the most", "knew what to do", or "knew the answers". This was often the definitive quality of scientists – their expert knowledge and behavior. In addition, viewers were especially drawn to other young people, both in dramatic roles and documentary films, who were striving to demonstrate competence. One girl's comment about a *3-2-1 Contact* cast member's venturing into a salt marsh captures this feeling: "Trini was doing a lot of things she's never done before. She was pretty brave to do 'em. Most girls wouldn't do that!"

Presentation of scientists

In their essays about scientists and responses to specific television portrayals, children held both positive and negative perceptions of scientists and the work of science. Over two decades ago, Mead and Metraux (1957) found a similar ambivalence toward scientists among high school students.

On the positive side, our target audience respected scientists for their expertise and contributions to humanity, especially for finding new cures for disease and inventing new things. However, many children did not see science as a personal or career interest. They saw it as intellectually difficult, requiring long training, and physically dangerous. Scientists were viewed as narrow human beings who spend their lives in laboratories and have little family or social life.

When children did say they wanted to become scientists, it was frequently to learn more about the human body. As one 9-year-old boy wrote, "I would like experimenting with the brain and other organs of the body because sometimes I think how marvellous we really are".

Conclusion

A fundamental issue faced throughout the development of the series was the presentation of science in ways that would maximise its appeal with children. In the early stages of the project, it was not clear whether the phenomena of science would make for compelling viewing by themselves or whether the commercial entertainment techniques of television, such as songs, comedy sketches, and stylised characters, would be needed to bolster the appeal of the content.

Formative research was able to illuminate this problem. Testing indicated that children appreciated the realistic, sometimes serious, tone of many documentary films and that their eagerness to seek out information about the world around them coexisted with an appetite for dramatic conflict, fast-paced action, and situation comedy.

This finding led us to conclude that a science series could be built around young people conducting an earnest investigation of interesting phenomena, and this tone was ultimately adopted by *3-2-1 Contact*. In making this recommendation, formative research provided child-based support for a view shared by our staff scientists: that science content need not be minimised in order to be entertaining.

"Child-based" is the key insight into our formative research. Through formative research, the target audience child has an in-house advocate and a voice at the decision-making table. Both the process and the product of formative research have been valuable in developing *3-2-1 Contact*. Both should be similarly useful to others who seek to combine entertainment and educational values in producing material for children.

Acknowledgements

Portions of this chapter are based on previous reports by the authors (Mielke and Chen, 1980, 1981). The latter work was supported by National Science Foundation Award No. 8020774. Any opinions, findings, conclusions or recommendations expressed herein are those of the authors, and do not necessarily reflect the views of the National Science Foundation.

References

Chen, M. (1980–81). Television, science and children: formative evaluation for *3-2-1 Contact*. *Journal of Educational Technology Systems* **9**, 261–276

Mead, M. and Metraux, R. (1957). Image of the scientist among high school students. *Science* **126**, 384–390

Mielke, K. and Chen, M. (1980). Making contact: formative research in touch with children. *CTW International Research Notes*. Children's Television Workshop, New York

Mielke, K. and Chen, M. (1981). *Children, television, and science: an overview of the formative research for* 3-2-1 Contact. Children's Television Workshop, New York

Mielke, K., Chen, M., Clarke, H. and Myerson Katz, B. (1978). *Survey of television viewing interests among eight-to-twelve-year-olds*. Children's Television Workshop, New York

Myerson Katz, B. (1980). CTW's new science series: the role of formative research. *Television* **7**, 24–31

Scriven, M. (1980). *Evaluation Thesaurus*. Edgepress, Point Reyes, California

Adult Learning from Educational Television: The Open University Experience

Anthony W. Bates

The importance of context

I am often asked by visitors to the Open University: "Is television as educationally effective as books?" It is an impossible question to answer, because it is based on certain premises which themselves need to be questioned. The question assumes that the teaching function of television would be the same as that of books; it also assumes that it is necessary or desirable to choose between television and books, rather than use them together. Above all, the question can only be answered within a given context.

Ten years' research at the Open University into students' learning from television indicates very clearly that the extent to which students will learn from television depends entirely on the conditions surrounding the use of television. What might work under laboratory conditions will be very different if a student has to view at 6.00 am, is behind with his assignment, and does not understand something in the course that is not covered by the programme. Context then is the key to understanding students' learning from television, in any circumstances, but nowhere is this more crucial than in the field of adult education.

Consequently, I make no apology for concentrating this paper on adults learning from television in the context of the Open University in Britain. With over 80 000 students enrolled in 1982, over 40 000 graduates since the first students enrolled in 1971, and over 1500 television programmes broadcast each year, the Open University must be one of the world's major users of television for adult education. While its unique nature means that generalisations to other contexts where adults learn from television cannot lightly be made, we have learned much about what affects whether or not students *will* learn from television – and more importantly, *what* they learn from television.

The Open University was founded in 1969 to provide opportunities in higher education for adults in Britain. It is based on the assumption that most of its students will be in full-time employment, and will therefore need to study primarily in their spare time, at home. Enrolment is open to any British resident over the age of 20, irrespective of previous educational qualifications, who is prepared to put in the time and effort required to study. Courses are specially prepared by academics employed full-time by the Open University, which is an autonomous, degree-awarding institution, with the same status as other British Universities. It is directly funded by the Government, with a budget in 1982 of over £50 million.

Students use a combination of specially printed correspondence texts, television, radio or audio-cassettes, and for some Science and Technology courses, home experiment kits. There are also, for some courses, compulsory one-week summer schools, and face-to-face tuition is available at local centres for the larger courses. Each student has a correspondence tutor, who both marks assignments and advises the student. Students gain credits through the combination of tutor- and computer-marked assignments during the course, and an end of course examination, which is supervised. Six credits are needed for an ordinary degree, and eight for an honours. A credit requires approximately 12 hours per week of study over a 32 week period, or roughly 400 hours minimum per credit. There are also courses which can be taken individually, distinct from the full degree programme. These continuing education courses may be professional up-dating, consumer or community education, etc.

For most courses, the "core" of the course is the correspondence texts, sometimes supplemented by a set book or readers (collections of articles). At most, students will get one 25-minute television programme a week (32 in a full course); sometimes, they may get only one television programme a month. There are a few courses (about 5%) with no television programmes at all.

The Open University has a unique partnership with the BBC. The BBC produces television, radio and audio-cassette material in conjunction with Open University academic and other course design staff. Producers are full members of course teams. The Government, through its grant to the University, pays the full cost of the BBC's services to the University. The BBC has set up a department (BBC/OUP) specifically to provide services to the Open University. In Autumn 1981, a brand-new, £5½ million purpose-built studio complex became operational on the University campus, and is managed by the BBC on the University's behalf. Currently, over 1500 Open University television programmes are transmitted each year on the BBC's national television networks, requiring over 35 hours a week of transmission time.

This is the context in which we have studied adult learning from television. Adult learning in this context is independent, self-motivational, and relatively isolated. It involves considerable personal sacrifice, both financially and particularly in terms of leisure time, but with a promise of future rewards, financial in some cases, but more often involving improved self-concept. However, we are as much concerned with the differences between learning from television and learning from other media, as we are in differences in learning between adults and children. It is necessary to provide a theoretical framework into which the role of television in learning can be placed, and this means looking at the nature of television as a medium for teaching adults.

Educational characteristics of television

Heidt (1978) provides a useful critique of various approaches to media selection and discusses differences between media. However, none of the theories of media selection discussed by Heidt really attack the problem of the relationship between media and learning. The most relevant work in this area is clearly that of Salomon (1979), but apart from some convincing research on the relationship between television and the development of perceptual skills, even Salomon's work is at best vague about the relationship between specific media and the process of learning and thinking. There seems to be so little research or literature on this issue that one is tempted to ask: *Is* the process of learning from television different from learning from other media? This question is also addressed in Chapter 5 of this book. It is essential then to be clear about the educational function of television, when considering learning from it.

The first thing that needs to be done is to differentiate between three broad types of characteristics of television which are often confused. These are:

Type 1: the distributional and social characteristics of television.
Type 2: the control characteristics of television.
Type 3: the symbolic (or audio-visual) characteristics of television.

TYPE 1: DISTRIBUTIONAL AND SOCIAL CHARACTERISTICS

With regard to adults, one of the most valuable characteristics of television is the access it offers to potential learners. In the major developed countries there is a set in practically every home. There is much unplanned or incidental viewing, and this enables educational programmes or messages to reach many people otherwise out of reach. Television, and the link with the BBC, was vital in establishing the image of the Open University in the early days, and through this, in recruiting students.

There are also certain characteristics of general television that carry over to specifically educational programmes. It is a familiar medium to the user, convenient and easy to use, and is usually associated with pleasurable experience, through its power to attract and entertain. These characteristics are extremely important for nonformal or adult education, and have been well exploited by series such as the BBC's *On the Move*, as part of the Adult Literacy campaign. Thus, even if there were no differences in the process of learning from television or books, television would still have an important educational role, through its power to reach specific adult audiences. However, there are additional characteristics of television, besides its accessibility and liveliness, which are important from an educational point of view.

TYPE 2: CONTROL CHARACTERISTICS

Whereas characteristics of the first type distinguish television (and radio) from other media, control characteristics vary *within* the medium of television. They are not necessarily unique to television, but may also apply to other media. However, much confusion is caused by failing to separate the control characteristics of television from other aspects.

These characteristics reflect the extent of control the learner has over when and where he or she uses the medium, or even over *how* the medium is used. Although these features could be elaborated a little more, I make the distinction between three types of television which have very different control features, each of which affects learning in quite different ways. These are broadcast television, recorded television, and television with "added-on" features.

Broadcast television. Table 1 lists the control features associated with broadcast television, and the implications for learners. Some of the implications for learners in Table 1 are obvious, but some need more comment. Because it is necessary to watch at fixed times, and because it is not possible to stop or interrupt a broadcast at a specific point, it is more difficult for the learner to integrate or relate broadcast materials to other learning. If ideas or thoughts are stimulated *during* a programme, the learner runs the risk of either losing the thread of the programme, or the thread of his or her ideas, unless there is a mental capacity for doing both

Table 1   Control characteristics of broadcast television

TV characteristic	Learner implications
Fixed schedules	Fixed time to view
Scarcity of time (hence only one or two transmissions)	Limited response to material
Ephemeral	Nonrepeatable; nonretrievable (except by memory)
Continuous	Thinking "on-the-run"
Holistic (i.e. a single unit)	Reflexion, analysis, restructuring, relating to other materials, all difficult
Aimed at "average" target viewer	No room for individual differences in pace
Programme pace, level, format, structure, decided by broadcasters	Dependency on "responsible" broadcasting

simultaneously. Of course, this is not a unique problem – anyone who sits through a formal lecture has the same difficulty, but it should be noted that with a television programme, the range and amount of information to be processed is much greater, being multichannel. Some producers at the Open University have argued that broadcasting teaches learners to think "on the run", and that this is an essential everyday skill.

Lectures can be retrieved to some extent by notes, which use the same verbal coding system as the medium in which the information is transmitted. With television, however, visual information in particular is difficult to code in note form, particularly since the coding of the overall "meaning" of the total experience is done verbally, in the form of written notes. Our research at the Open University has shown that most students find it impossible to take notes while viewing, and those that do are usually very dissatisfied with their notes.

No matter what the level of programme, the very nature of broadcasting means that the target audience will vary considerably in language skills and general ability. Programme makers have to make assumptions about the "appropriate" level, but there will always be a majority of the target audience who will not find the pace suitable, in terms of strict learning goals. The continuous and fixed pace of a programme does not allow for individuals to rework, or "jump" in their thinking to the level which best suits them. However, this limitation of broadcast television can be offset to some extent by the diversity of interpretations or levels of thinking that certain kinds of programming permit.

The fact that broadcasts reach the learner as a complete package, made by professional broadcasters who cannot be interrogated by the learner regarding the selection, editing and structuring of material, has two major implications. It makes the learner dependent on "responsible" broadcasting, and regrettably, this in turn tends to encourage an "accepting" level of response from the learner. (That this *need* not be so does not make the point any less valid.)

For these reasons, specifically, educational programming tends to be accompanied by support materials, in the form of teachers' notes and follow-up exercises and activities.

Recorded television. Table 2 lists the control features associated with recorded television, and the implications for learners. The advantages to learners of recorded material over broadcast material are well known, and generally need little elaboration. It is possible, if learners have the time, and the level of the material is not too wide of the mark, for learners with a wide range of abilities to repeat the material until they have mastery of it. Course designers can integrate video material more closely with other learning materials, so learners can move between different media as and when required. More importantly, learners can control their learning, so that they can reflect on the material, analyse or restructure it, as best suits them. The ability to create "chunks" of learning material, or to edit and

Table 2   Control characteristics of recorded television

TV characteristics	Learner implications
Available when required	Convenient
Rewind/fast-forward facility	Repetition; mastery learning
Stop-start facility	Integration with other media; activities integrated with video; more room for individual variation
Hold-frame facility	Analysis of detail
Noncontinuous/segmented*	Reflexion, analysis, restructuring easier
Editing facility	More selectivity; more questioning

* If made for use on cassette only

reconstruct video material, can help develop a more questioning approach to the presentation of video material. Recorded television therefore considerably increases the control of the learner (and the teacher) over the way video material can be used for learning purposes.

It is worth pointing out that while the learner has *less* control over the learning situation while watching broadcast television than in a conventional classroom or lecture setting, in that he or she cannot interrupt, interrogate or ask for clarification, with recorded television the learner has *more* control. The learner also has the possibility of becoming much more independent. However, this depends to some extent on the *accessibility* of video replay equipment. The rapid increase in home ownership or rental of video machines will make independent learning from television a practical reality for adult and continuing education in the near future.

"Add-on features". I do not wish to elaborate on this section, but new developments in technology which use the standard domestic television set as a combined loudspeaker and visual display unit, such as Teletext (CEEFAX, ORACLE), viewdata systems (PRESTEL, TELEDON), electronic blackboards and sonically-coded visual systems stored on audiocassette (CYCLOPS), and computer assisted learning, all enable increased interaction and control by the independent learner. Television-related technology is moving towards increased learner control, greater possibilities for interaction between learner and teaching materials, or learner and teacher, and greater sophistication in the educational use of the domestic television set.

TYPE 3: SYMBOLIC CHARACTERISTICS

This set of characteristics is quite different from either distributional or control characteristics. In this set of characteristics, I am concerned with the $60 000 question: "Is there something *different* about information presented audio-visually through television, than information presented personally by a teacher, or through print, or through other media?" If so, what are the learning implications?

To answer that question, one must look at the role that media play in teaching. A medium is one way through which knowledge can be represented – it is a window on the world. Knowledge of the same topic or concept though can be represented in many ways: verbally, numerically, physically, conceptually, symbolically. Thus the experience of "heat" can be represented by words ("it is hot"), by numbers (110°F), physically by touch (feeling the heat), by concept ("form of energy arising from random motion of molecules of bodies"), or symbolically (a man dragging himself through the desert).

Each way of "knowing" heat is different. Eventually though, the learner must relate all these experiences together to fully understand the idea or "concept" of "heat". Teaching aims at finding suitable *means* of conveying or representing knowledge and skills to learners. Television may be *one* of those means. Thus television may provide a different way of knowing about a concept from being told about it or experiencing it. And teaching involves more than just presenting or representing knowledge of the world in a variety of ways. It also concerns *using* that knowledge.

Olson and Bruner (1974) have used this distinction between *acquiring* and *using* knowledge to argue that *knowledge* (or content) is invariant across media, but mental *skills* are more dependent on the right choice of medium for their development. Thus learners can acquire facts, ideas, principles, opinions, relationships from any medium, whereas skills such as observation, analysis, problem solving are developed better by some media rather than others.

A refinement of this hypothesis is Salomon's proposition (1979) that the symbol systems unique to different media do not facilitate learning in a simple, unidimensional manner, but can facilitate learning in one of three ways:

(*a*) by *activating* already existing mental skills, through providing practice in their use;

(*b*) by *short-circuiting* difficult mental processes, through symbol systems representing knowledge in a new way;

(*c*) by *supplanting or modelling* the mental elaborations required – i.e. demonstrating to learners *how* to move from point A to point B – to incorporate new knowledge.

Salomon also argues that if learners are already familiar or knowledgeable in a specific area, their response to the use of different media will be different than if the knowledge to be learned is new. If learners are already familiar with the subject area, choice of medium will be less important for knowledge acquisition. However, the more the medium presents knowledge which differs from the learner's already existing knowledge base, the more mental effort or elaboration will be required to incorporate that knowledge. In such circumstances, it *will* matter how well the medium is matched to the learning task.

He goes on to argue that where media enable difficult mental processes to be short-circuited, knowledge acquisition is speeded up, but mental skills are not developed. Modelling, on the other hand, will develop mental skills, if these do not exist or are not well-developed, but modelling will actually interfere with learning if the learner is already knowledgeable in the area.

I have considerable reservations about these hypotheses (see Bates, 1981, for a full critique), but they are useful in bringing to attention the nature of learning through media, and there is some support for such hypotheses from the Open University's own experience. For instance, although it may not always be conscious policy, in Open University courses the same content is usually dealt with in different media, but different approaches or ways of dealing with that common content are found in each medium, resulting in students processing or using that knowledge in different ways.

Thus a medium has *two* functions in the learning process:

(*a*) the presentation of knowledge in a different way from the presentation of the same knowledge through another medium, thus providing a broader base of "knowing" – knowing "what" in different ways;

(*b*) the development of certain mental skills in *using* knowledge – knowing "how" in different ways.

If such a theory has a strong empirical base (and this has yet to be demonstrated), it is necessary for teachers and producers to identify clearly those skills which are most suited to development through particular media. I hope to show that some of our research can be seen as identifying some such skills.

What roles does television play in Open University learning?

The University has, over the past 12 years, jointly produced with the BBC over 3000 television programmes and a similar number of radio programmes, plus over 500 cassettes and 100 records. The range and variety of output has been enormous. It could be argued that each programme is a unique "event". Any attempt at classification is extremely difficult.

With such a wide range of possible uses my group has had to concentrate its more detailed research into student learning on specific examples of programme types or functions which are commonly used in the Open University, or represent possibly major new developments in the use of audio-visual media. Specifically, we have looked at television used for the following purposes:

the presentation of abstract mathematical concepts through silent animation (Mathematics 231 – Ahrens, Burt and Gallagher, 1975);

the construction of physical models to represent abstract ideas (Technology 291 – Bates, 1975a);

the illustration, through a location visit, of industrial applications of chemical processes (Science 24 – Gallagher, 1975a);

the use of dramatisation for enriching students' interpretation of a novel (Arts 302 – Brown and Gallagher, 1978);

the reinforcement of techniques and concepts dealt with in other components (Science 333 – Berrigan, 1976; Social Studies 101 – Kern, 1976a);

the development of skills in using television as part of Open University studies (Social Studies 302 – Gallagher, 1977a; Social Studies 101 – Morgan, 1978);

the use of television for presenting case-study material (Educational Studies 221 – Gallagher, 1977b; Social Studies 282 – Brahmawong and Bates, 1977; Technology 101 – Brown, 1981);

the effect of co-production on the learning effectiveness of programmes for Open University students (Science 354 – Marcus, 1980).

In this chapter I cannot deal fully with the question of the teaching roles of television in the Open University. However, it is of relevance to the theme of student learning that an analysis of the programme and other course material for those programmes that we studied showed that in general, the programmes *were* providing knowledge or developing skills in ways that were not found elsewhere in the course.

However, the fact that the programmes were being used in this way did not itself guarantee that the required learning would take place. For nearly every different teaching function examined, we have found examples where the strategy of using television has been both successful and unsuccessful. In other words, there are *other* conditions which have to be met as well.

To provide examples of this, I will choose two very different kinds of programmes, television case-studies, and programmes which aim to reinforce concepts taught elsewhere in the course.

Case-studies and documentaries

The use of television to provide case-study material is very widespread in the Open University. We have made several detailed studies of such programmes. A full summary of results, a critique, and advice to academics involved in such programmes, can be found in Bates and Gallagher (1977). There is also an excellent discussion of this kind of Open University programming by Grahame Thompson of the University's Social Science Faculty (Thompson, 1979). The main results we found were as follows:

1. The *functions* of the programmes, in terms of the skills required of students and the way content was treated, were quite different from those of the correspondence texts. The correspondence texts were theoretical, analytical and didactic. The television programmes dealt with concrete situations, presented "images" of complex, real-world situations, and were open-ended, open to interpretation, and nonanalytical. Producers and academics expected students to *analyse* the television material, using the theoretical or analytic constructs provided in the correspondence texts; to *apply* what they had learned in the texts to the real-world situations observed in the television programmes; to *generalise* or *draw conclusions* from the specific instances in the programmes; to *test, evaluate,* or *compare* the applicability of general principles in the text to the "real-world" instances found in the television programmes.

2. In several instances, the programmes were the *only* place in the course where students were able to develop or practice these kinds of learning skills, other than in answering assignment or examination questions. (In some cases, "readers" – collections of papers – or written case-studies provided a similar opportunity).

3. Academics and producers in most cases agreed on the relevance of the programmes to the course – they saw a congruence between the content of the programmes and the content of the correspondence texts, although the *treatment* of content and the choice of *examples* were different.

4. For some of the programmes investigated, a majority of the students did not think that the programmes were very relevant or helpful ("I learned nothing new from this.").

5. For most of the programmes investigated, a large minority (about a third) of students misunderstood the purpose or *function* of the programmes – they were looking to the programmes to provide *new content*, or *explanation* of difficulties encountered in the correspondence texts.

6. Less than one-third of students, for any programme, both understood the purpose of such programmes *and* appeared able to use the programme material in the ways suggested in (1) above. One-third of students on average did understand the purpose of such programmes, but appeared *unable* to use the programme material in the way intended.

7. A majority of students wanted more help in using such programmes.

It seems clear from these particular kinds of programmes that television is being used to encourage high-level mental skills (high-level, in the sense that they depend on the development of other learning skills, before they can be used – see Bloom, 1956, or Gagné, 1970). It is also clear that many students, even at third level, with many Open University courses behind them, find it difficult to use television in this way. Students' *expectations* of the role of media such as television are also different – they tend to judge programme material by the extent to which it adds new knowledge or explains knowledge inadequately dealt with in the texts; not by the extent to which it enables them to *use* the knowledge gained in texts.

The implication is quite clear. Students do *not* automatically know *how* to use instructional television to the best advantage. The further a programme moves away from overtly didactic teaching, the more help students need to develop the necessary skills to benefit from such programmes.

Television as reinforcement

In contrast to case-study programmes, Social Studies 102, TV programme number 7 (TV7), was intended to directly reinforce a section of the Unit 7 correspondence text. Thus both the programmes and the text dealt directly with "Hockett's Design Features". Both the producer and the academic responsible for the unit believed that a full understanding of the 16 design features described in the text required both vision and motion to show the concepts in action. A feature of the programme was the participative exercise, where six different film sequences illustrating instances of communication between animals or humans were shown to the students, who were asked to select during the programme the correct design feature which each sequence illustrated. An evaluation of the programme by Kern (1976a) showed that the programme was very highly rated by the students, who found the programme helpful and interesting. A majority of students stated that the programme gave them a better understanding of the 16 features.

Science 333, TV3 had similar aims of reinforcement of material covered elsewhere in the course. Two sections in the Techniques Handbook which were considered essential but difficult were also covered in the television programme. Although an evaluation (Berrigan, 1976) showed that generally students reacted favourably to the programme, a higher proportion of students (about one-third in all) was dissatisfied with the programme than was the case with Social Studies 101, TV7. There were several reasons for this. The main student criticism of Science 333, TV3 was that the programme did not deal *adequately* with the difficulties encountered in the Techniques Handbook. The two sections specifically dealt with were not treated in sufficient depth to help students who had found difficulties with these sections; other sections in the Handbook which caused some students more difficulty were not dealt with at all. While Social Studies 101, TV7 had been tried out in advance on a small group of students to get the pace right, no attempt was made on Science 333, TV3 to ascertain exactly what the difficulties were, and how best to treat them.

The general point underlying the research into the use of television both for case-studies and for direct reinforcement of textual material is that both uses of the medium can be valuable but the necessary conditions for success have to be met. I shall therefore in the rest of this paper try to identify what these main conditions are in the Open University context, and how these conditions affect learning from television.

Delivery

The first condition that has to be met in a distance learning system is the *delivery* of material. In the Open University this is a particular problem for television and radio. In 1974, when all courses had repeats, and at least one transmission at times which the students themselves rated as convenient, the mean viewing rate across all programmes and courses was 64%. This can be interpreted in one of two ways: any individual student was likely to watch almost two-thirds of the programmes on his course; or any individual programme was likely to be seen by almost two-thirds of the students on that course (Bates, 1975a). This though is an average figure covering the whole year. Viewing drops off considerably during the summer period, due to holidays and summer schools, and at the end of the academic year, as students concentrate more on revision for exams (Bates, 1975a). Up to about July, viewing figures for most programmes were averaging between 75%–80% in 1974, while the figure for the whole year was 64%.

Five years later, the average viewing figure was around 55%, a drop of over 8% (Grundin, 1980). The reduction was not spread evenly between the faculties. Science courses (which in 1974 had the highest average viewing figures) dropped 13%, whereas Mathematics courses (which had the lowest viewing rates in 1974) hardly dropped at all. Close examination of the data shows that the drop in viewing rates can almost all be explained in terms of deteriorating transmission times, for two reasons. The same courses surveyed in 1974 or 1976 and then again in 1979 (and there were 26 such courses) had dropped their viewing rate by an average of 11% although the course and its programmes were the same. Secondly, new courses in their first year of presentation, even in 1979, were given generally good viewing times and maintained on average good viewing figures in that first year of presentation.

The deterioration in transmission times is due to two factors. The first is the gradual deterioration in the overall mean "quality" of transmission time – more times being used which are less convenient to students. The second is the removal of repeat transmissions, which reduces the choice students have of times at which to watch.

Furthermore, the situation has deteriorated even further in 1982, when evening transmissions were lost. Our predictions (which have been extremely accurate over the last five years) are that the mean viewing rate will have dropped below 50% in 1982 (although a proposed video-cassette loan scheme may just keep it at around 50%). If the average viewing opportunity in 1982 had been the same as in 1974, there would have been 180 000 more student viewings of Open University television programmes than what we predict will have occurred in 1982. This is a tremendous loss of viewing. If students do not see programmes, they cannot learn from them. No single factor can have had a greater impact on student learning from television than this loss of viewing.

If the mean quality of transmission time cannot be maintained, effective learning from television for the *majority* of Open University students can be achieved only by drastically reducing the number of programmes for transmission (to give more repeats), or by finding additional means of viewing (such as through a video-cassette scheme).

While the transmission problem is very serious, it needs to be kept in proportion. Many courses will continue to receive good transmission times on a Saturday or Sunday, and even in 1982, many programmes were still repeated or also available on cassette. So what can be done to ensure that students who are able to watch can learn effectively from them?

"Relevance"

One of the main reasons given by students for not watching or listening to Open University broadcasts – or not treating them seriously – is their perceived lack of "relevance". There are two main reasons why programmes are not perceived as relevant by students. Some of these programmes *are* relevant, but students do not have sufficient information or skills to *see* the relevance of the programme material. In other cases, the material is *not* relevant, in that, with the best will in the world, even the course team would find it difficult to make links between the television programme and the rest of the course.

I shall not deal with the latter case. In none of our evaluations have we so far come across programmes that were not relevant to the main aims and objectives of the course. The question then is even more curious: why do so many students so often *perceive* programmes to be irrelevant? For instance, Brown (1981), in his evaluation of Technology 101 broadcasts, found that although students often learned what the course teams wanted them to learn from the programmes, they were not only unable to see the relevance or significance of the programmes, but they were also unaware that they had learned something important from the programmes. This would not matter, of course, if it had not influenced their viewing behaviour. However, unfortunately, after a good start, viewing figures by the end of the year were lower on Technology 101 than on most other foundation courses. Lack of relevance was one of the main reasons given by students for nonviewing.

Perceived relevance is a complex phenomenon. It seems to be influenced by several different factors: assessment policy; overt integration of broadcasts with texts; timing of broadcasts in relation to study of other components; scheduling of print and broadcast productions; and choice of inappropriate programme material. I will deal with each of these in turn.

Assessment policy. Anything included in a course by a course team – including the programmes – might be considered relevant by definition. However, students have a very instrumental approach to studying at the Open University. Time is a precious commodity, so for many, their main criterion for judging relevance is: will it help me get better grades? Put at its crudest, this can be rephrased as: can I get good marks on assignment and examination questions without watching or listening to the broadcasts?

This presents a problem for the broadcasts. With transmission times getting more difficult, course teams are not allowed to set questions which can be answered only if students have seen or heard a programme, unless an alternative question is set. One study (Gallagher, 1977b) showed that only one-fifth of the students attempted a broadcast-related question (although 85% of students saw the programme), and the average grades awarded for the broadcast-related assignment was below that for the alternative assignment questions. There is evidence that deliberately setting computer-marked assignment questions which can only be answered through watching the programmes does improve helpfulness, relevance and viewing figures (Bates, 1973; Bates, 1975c; Kern, 1977a) but also increases student resentment and can lead to trivial questions, or questions which do not really test the essential audio-visual component of the programmes (Bates, 1973; Kern, 1977b). Also if the role of media is to present (the same) knowledge in new ways, or to develop certain skills, such an approach seems to be the wrong way to tackle the problem. Thus *overtly* setting assignments on broadcasts may improve the perceived relevance of programmes, but possibly at the expense of a sensible assessment strategy, and may disadvantage students unnecessarily (see Kern, 1976b and Bates, 1975c for a full discussion of this issue).

The main difficulty appears to be how to help students see the relevance of material and approaches contained in the broadcasts, and to draw on this – and other – material when tackling assignments. The key to this lies in the integration between text and programme material (see below), but one factor which would help would be a clear recognition and identification of the skills which are *specifically* intended to be developed through broadcasts (if any). For instance, given that the intention of many of the case-study programmes we have analysed has been to develop skills of analysis, application, generalisation, synthesis, etc., it would seem perfectly valid to set assignment questions which deliberately attempted to test these skills. If students are clear at the beginning of the course that this is one of the roles for broadcasting, they might see the relevance of the programmes in a different light. It might also be noted that a fair test of the development of such skills might be a request to students to analyse a short programme segment – on video-cassette – under examination conditions (e.g. "Explain the situation portrayed in the programme segment in terms of the main principles covered in the correspondence texts."). In order to be able to do this, however, course teams must be clear about what skills they intend television, radio or cassettes to develop, as distinct from the texts, and these intentions must be successfully communicated to students through the programmes themselves.

Integration of broadcasts with texts. A main difficulty for many students is the lack of explicit links between the programmes and the correspondence texts. It is the exception rather than the rule to see references to television or radio programmes in the text – and even rarer to find in the text any form of analysis or discussion of programme material. Conversely, it is rare for programmes to refer explicitly to sections of the text. (This is often due to the different production schedules for text and broadcasts – see below.) Frequently, the bridge between text and programme is made through the broadcast notes, and it is not surprising then that on many courses, the broadcast notes are rated more helpful than the programmes themselves. It is clear from many of the studies (e.g. Gallagher, 1975b; Gallagher, 1977b; Brahmawong and Bates, 1977; Brown, 1981) that students find it very difficult to make the connection between television and the text, even though a great deal of help may be given in the broadcast notes. The success of Mathematics 101 in particular, where texts and broadcasts have been tightly integrated, and of audio-cassettes, where again the links between texts and programmes are explicit, indicates the importance to students of explicit links contained in both programmes and text. The need for these explicit links is *not* because students are stupid or lazy. In discussion with producers and academics, we have found that it is often extremely difficult for *them* to explain the conceptual relationship between programme and text. For *students* to do this without guidance, a full understanding of the subject area, and high-level skills, are required. There will always be some who can do this – but there will be many others who cannot, unless helped.

Timing of broadcasts in relation to other components. Some of the earlier studies carried out by researchers at the Open University (Ahrens *et al.*, 1975; Bates, 1975b; Gallagher, 1975b) indicated that most students were usually studying the texts at least two or three weeks behind schedule, which meant that broadcasts were often watched or listened to *before* students had covered those parts of the correspondence texts to which programmes were linked (or were missed altogether, because students were trying to catch up). These studies also established that students' work patterns were determined mainly by the tutor-marked assignment cut-off dates rather than by the timing of the broadcasts. This meant that programmes were often seen considerably out of synchronisation with study of the textual material, and this obviously affected the perceived relevance and helpfulness of the programmes. A number of suggestions have been made – that programmes should be introductory rather than depend on prior reading of the texts, or that repeats should be shown three weeks after transmission, etc.

However, rather than accept these compromises on integration, the Mathematics 101 course team took the bull by the horns, and structured the television, radio and audio-cassette material so tightly with the text that great pressure was put on the students to work to schedule on *all* components. There were sections in the text which directed students to watch or listen to programmes, then carry out work (in the text itself) based on the programmes.

This strategy was highly successful. A study by Womphrey (1978), showed that over 75% of responding Mathematics 101 students kept up with or in advance of the television pacing, and of those who were behind the television pacing, half were less than one week behind. The Mathematics 101 viewing rate was 78% (compared with 63% for Mathematics 100), and the helpfulness ratings for Mathematics 101 were much higher than for Mathematics 100, and the second highest in the faculty.

It is an example where close integration of programmes and text overcame the scheduling problem and increased the relevance of at least the television programmes.

Scheduling of print and programme production. To obtain such close integration between text and programmes, the schedule of print and programme production has to be carefully synchronised, so that programme scripting, editing and commentary can take account of draft texts, and final drafts of texts can take account of programme material. In practice, however, it has often proved extremely difficult to obtain such close synchronisation of production schedules. Some course teams have not had producers available or assigned in time. On other course teams, key academics have not been available – or their draft texts have not been written – when programmes were to be scripted or edited. On some course teams, no attempt has been made to synchronise broadcast and text production, because, due to prior staff commitments, they were on different time schedules from the beginning.

Lack of synchronisation in production is probably the main reason why it is difficult to make explicit links between programmes and texts. I do not wish to minimise the difficulties of two large and complex organisations, both with highly individual and creative staff, achieving such detailed synchronisation of activities. The will is usually there, but practical difficulties often render synchronisation impossible. Even then, as I have previously noted, programmes are usually relevant to the aims and goals of the course. However, when students need to make conceptual leaps to link programme material with texts, with little help other than from broadcast notes, there is often a major loss of learning from the programmes. In such cases, the programmes are wrongly seen by students as irrelevant, and hence dispensable.

Inappropriate programme material. Although it was not a frequently occurring event in our evaluations, the choice of material which was visually interesting or attractive, but not directly relevant to the aims or purposes of the programme, sometimes drew vehement criticism from students. For instance, in Science 333, TV3 (Berrigan, 1976) students found a section which showed a volcano erupting extremely irritating, as it threw no light on the topics under discussion, and seemed to have been included at the expense of some of the more relevant and difficult aspects of the topics under consideration. However, it must be said that this appears to be a relatively infrequent and minor factor affecting students' judgment of relevance. Much more important are the structural relationships between texts and programmes.

Individual differences

There appear to be strong individual differences between students in their ability to learn from different media. Research into radio at the Open University (Brown, 1980) showed that students vary enormously in their use of different media even before they began their studies. Students' response to Open University radio in particular is extremely varied. Whereas most students on most courses do seem to make an effort to watch the majority of television programmes, there is no coherent pattern for radio. Some students listen to none. Others listen to them all. Some listen to half. Students are equally distributed in their radio listening pattern (Bates, 1975a). I interpret this as an indication of strong individual preferences for learning from specific media.

There seems to be three conclusions one can draw. Although radio may not be used a great deal by a lot of students, those students who do use it regularly find it extremely valuable. (This is supported by Durbridge's study, 1980, of the use of radio on Science 101.) Secondly, given such wide individual differences in media preferences, it is important to provide a wide range of media for students. In this way, each student is likely to find at least one means of acquiring knowledge with which he or she is comfortable. Some redundancy in media provision may therefore be necessary, if dropout is to be kept down. Thirdly, it may be important to try to help students to make more use of less preferred media, through training and help within courses themselves.

One very important difference was noted in three different large-scale broadcast surveys (Bates, 1975a; Gallagher, 1977c; Grundin, 1978). In all these surveys, there was a greater tendency for students who passed their exams to make use of broadcasts *more* than students who dropped out or failed their exam; and students with grades A or B watched or listened more to Open University programmes than students with pass grades C or D. These results are not surprising, and are probably more to do with students' motivation and workload, although it is clear evidence that the broadcasts do not have *negative* effects. What is particularly significant however, is that the *weaker* students who do watch or listen rate the broadcasts as *more* helpful than do the more successful students, and this trend is consistent from fail up to A/B – a trend opposite to that for correspondence texts. Having examined other possible explanations, Grundin (1978) concludes:

It is possible though that broadcasts in many courses are genuinely more helpful for weak students than for more successful students. Since weak students tend to find the correspondence texts difficult and less helpful (than strong students), these students would seem to need more help from other components. (p. 24)

It therefore seems that students who are satisfied with the texts will find the programmes less helpful, since they do not need further explanation. Note how closely these conclusions support Salomon's theory of the differential effects of learning through media (p. 10). Note also the danger of concluding that audio-visual media are not necessary because successful students do not need them.

There are important student differences identified by the research. Gallagher (1977b) found three kinds of student "types" in response to open-ended documentary-style television programmes: those that wanted straight instruction ("didacticists"); those that appreciated the value of open-ended programmes, but wanted more help ("guidance-seekers"); and those that were happy with open-ended programmes ("explorers").

Again, it would seem important to find ways of helping students to broaden their learning styles, through help in the programmes themselves, so that their approach to learning matches the task.

Broadcast notes

A great deal of research has been done on broadcast notes (see particularly Kern, 1976c). It can be seen that unless programmes are tightly integrated with texts as in Mathematics 101, courses will usually need some form of broadcast notes. Technology 101 in its first year of presentation tried to avoid broadcast notes, first of all by attempting to get the texts to comment on the programmes. When, due to scheduling difficulties, this proved to be impossible, a two-minute prologue was used to introduce each programme and relate it to the texts. This apparently was not a good substitute for broadcast notes, as students found it difficult to remember the points made in the prologue when watching the rest of the programme, and afterwards, they had no permanent record of the programme (Brown, 1981).

Students rarely read broadcast notes in detail before a broadcast, so any prereading should be kept to a minimum – usually a statement of objectives, and two or three main points to watch out for during the programme. Students do find a brief resumé of the main points of the programme, and diagrams or tables shown in the programme useful for revision purposes. If a lot of detail is provided, though, students will either not read the notes, or will read the notes, but not bother to watch or listen to the programme. The main follow-up activity that students will do after a programme is further reading – and only about a half at a maximum will do this. If follow-up activities are important, they are better incorporated in the main text, or included in cassette activities. The most common mistake with broadcast notes is to entrust them with the main task of detailed integration of the programme with the text. Some broadcast notes are over 40 pages long, and there is no way students can cope with that extent of verbal explanation and guidance on a single television programme (Gallagher, 1977b).

Conclusion

I have tried to cover a large body of research into learning from television at the Open University. I have not been able to discuss methodology at all, although in our evaluations of individual programmes, we try, through both quantitative and qualitative methods, to measure exactly what students have learned from their programmes, and how they have integrated this with their other learning. This requires analysis of the content of television programmes and the other materials in the course. The research does indicate that there are major difficulties that course designers face in getting adults to learn from television, in an integrated, multimedia distance teaching system such as the Open University. However, I hope I have been able to show that the research findings do in general suggest that television does have an important role in assisting adult learning at the Open University.

Acknowledgements

This chapter is based on two articles previously published: Some unique characteristics of television and some implications for teaching and learning. *Journal of Educational Television and Other Media* 7, No. 3, 1981; and "Learning from Audio-Visual Media: the Open University Experience", *Teaching at a Distance, Research Supplement No. 1*, 1982. I am grateful for permission to reproduce material from these articles.

References

Ahrens, S., Burt, G. and Gallagher, M. (1975). "Broadcast Evaluation Report No. 1: M231, 'Analysis' ". Mimeo, 42 pp. + appendix. Open University, Milton Keynes

Bates, A. W. (1973). "An Evaluation of the Effect of Basing an Assignment on Broadcast Material in a Multi-Media Course". *Programmed Learning and Educational Technology* **10**, No. 4, 18 pp. + appendix

Bates, A. W. (1975a). *Student Use of Open University Broadcasting.* Mimeo, 79 pp. + appendix. Open University, Milton Keynes

Bates, A. W. (1975b). "Broadcast Evaluation Report No. 3: T291, TV6 'Instrumentation' ". Mimeo, 37 pp. + appendix. Open University, Milton Keynes

Bates, A. W. (1975c). "Should Assignments be set on Broadcast Components?" A Paper to Exams and Assessment Committee. Mimeo, 4 pp. Open University, Milton Keynes

Bates, A. W. (1981). "Some unique characteristics of television and some implications for teaching and learning". *Journal of Educational Television and Other Media* **7**, No. 3, 79–86

Bates, A. W. and Gallagher, M. (1977). *Improving the Effectiveness of Open University Television Case-studies and Documentaries.* Mimeo, 40 pp. Open University, Milton Keynes

Berrigan, F. (1976). "Broadcast Evaluation Report No. 20: S333 'Earth Science Topics and Methods', TV3". Mimeo, 37 pp. + appendix. Open University, Milton Keynes

Bloom, B. S. (1956). *Taxonomy of Educational Objectives: the Classification of Educational Goals. Handbook 1: Cognitive Domain.* David McKay, New York

Brahmawong, C. and Bates, A. W. (1977). "Broadcast Evaluation Report No. 23: D232, 'National Income and Economic Policy', TV4". Mimeo, 23 pp. Open University, Milton Keynes

Brown, D. (1980). New students and radio at the Open University. *Educational Broadcasting International* **13**, No. 1

Brown, S. (1981). "T101 '2 + 6' Evaluation: Final Report on First Year Broadcast Presentation". Mimeo, first draft. Open University, Milton Keynes

Brown, D. and Gallagher, M. (1978). "Broadcast Evaluation Report No. 15: A302 'The Nineteenth Century Novel', TV9". Mimeo, 27 pp. Open University, Milton Keynes

Durbridge, N. (1980). *S101 Radio Programmes 1–9 and Audio-Cassettes 90 and 91.* Mimeo, 47 pp. + appendix. Open University, Milton Keynes

Gagné, R. M. (1970). *The Conditions of Learning.* Holt, Reinhart and Winston, New York

Gallagher, M. (1975a). "Broadcast Evaluation Report No. 4: S24, TV7". Mimeo, 46 pp. + appendix. Open University, Milton Keynes

Gallagher, M. (1975b). "Broadcast Evaluation Report No. 2: E221, 'Decision-Making in the British Education System', TV3, R6". Mimeo, 52 pp. + appendix. Open University, Milton Keynes

Gallagher, M. (1977a). "Broadcast Evaluation Report No. 23: D302, 'Patterns of Inequality', TV2". Mimeo, 23 pp. + appendix. Open University, Milton Keynes

Gallagher, M. (1977b). "Broadcast Evaluation Report No. 22: D302, 'Patterns of Inequality', TV1". Mimeo, 48 pp. + appendix. Open University, Milton Keynes

Gallagher, M. (1977c). *Broadcasting and the Open University Student.* Mimeo, 90 pp. + appendix. Open University, Milton Keynes

Grundin, H. (1978). *Broadcasting and the Open University Student: The 1977 Survey.* Mimeo, 35 pp. + appendix. Open University, Milton Keynes

Grundin, H. (1980). *Audio-Visual and Other Media in 91 Open University Courses: Results of 1979 Undergraduate Survey.* Mimeo, 45 pp. + appendix. Open University, Milton Keynes

Heidt, E. V. (1978). *Instructional Media and the Individual Learner.* Kogan Page, London

Kern, L. (1976a). "Broadcast Evaluation Report No. 21: D101, 'Making Sense of Society', TV7". Mimeo, 21 pp. + appendix. Open University, Milton Keynes

Kern, L. (1976b). *Basing Assignments on Broadcasts: A Paper to Exams and Assessment Committee.* 5 pp. Open University, Milton Keynes

Kern, L. (1976c). *Using Broadcast Notes in Distance Teaching.* Mimeo, 7 pp. Open University, Milton Keynes

Kern, L. (1977a). *Results of Students Reports on TD 342 Broadcasting Components*. Mimeo, 5 pp. + appendix. Open University, Milton Keynes

Kern, L. (1977b). "Student Feedback on Broadcast CMA Questions: An Interim Report on D204". Mimeo, 3 pp. + appendix. Open University, Milton Keynes

Marcus, R. (1980). *The Co-Production of the S354 Television Series*. Mimeo, 12 pp. Open University, Milton Keynes

Morgan, F. (1978). "Broadcast Evaluation Report No. 27: D101, 'Making Sense of Society', TV4". Mimeo, 67 pp. + appendix. Open University, Milton Keynes

Olson, D. and Bruner, J. (1974). Learning through experience and learning through media. In D. Olson (ed.). *Media and Symbols: The Forms of Expression*. The 73rd NSSE Yearbook, University of Chicago Press, Chicago

Salomon, G. (1979). *Interaction of Media, Cognition and Learning*. Jossey-Bass, London

Thompson, G. (1979). Television as text: Open University "case-study" programmes. *In* M. Barrett *et al.* (eds). *Ideology and Cultural Production*, 38 pp. Croom Helm, London

Womphrey, R. (1978) *M101 (1978) Feedback: Summary Results No. 3*. Mimeo, 24 pp. Open University, Milton Keynes

4

The Role of Television in the Formation of Children's Social Attitudes

Peter G. Christenson and Donald F. Roberts

Introduction

Public concern about how young people are influenced by the view of the world presented via whatever medium happens to be current is nothing new. The assumption that the values presented in symbolic materials will influence children's social attitudes can be seen in Plato's advocacy of censorship of storytellers, in outcries over print fiction as diverse as Dickens' serials, Grimm's Fairy Tales, and the penny westerns, and in more recent criticisms of US television content emanating from groups and organisations as varied as the National Organization of Women, the National Association for the Advancement of Colored People, and the Moral Majority. Two assumptions seem to underly this concern: (1) that children are particularly vulnerable because they lack the cognitive skills and the life experiences that enable adults more adequately to process such symbolic content; (2) that the fictional content of most media – but especially television – is particularly forceful, both because it combines representational realism and symbolic structure so as to focus attention and provide closure in ways that real life cannot do and, in the case of television, because of its ubiquity (see for example, Gerbner and Gross, 1976).

One cannot help but feel that much of the expressed concern also proceeds from an unspoken fear of any content that threatens the parental information monopoly (Roberts, 1974) or that fails to dovetail with the value and belief structures of specific interest groups. Indeed, even though recent efforts to use television for prosocial purposes have drawn attention to the possibility of using television to foster positive attitudes, public comment usually focuses on the "harmful" effects of television content, as exemplified in the many studies of television violence (e.g. Surgeon General's Scientific Advisory Committee, 1972), and some recent charges that television fosters sexist (Butler and Paisley, 1980) and racist (Pierce, 1980) attitudes.

But expressed public concern and charges that television strongly influences children's social attitudes, whether for good or ill, do not make it so. To determine whether and how television plays a role in the formation of children's attitudes and values, we must examine the empirical studies that have attempted to address this issue. This, in turn, requires a survey of a number of widely dispersed studies, dispersed in time, in location, and in the specific issues which they attend.

Scope of the chapter

Our general focus is on the relationship between television viewing and children's and adolescents' acquisition of social attitudes – norms, values, moral standards – with the exception of attitudes concerned with violence, an issue more fully covered elsewhere (chapter 7 in this volume; Comstock, Chaffee, Katzman, McCombs and Roberts, 1978). This exclusion still leaves a wide range of content, albeit much of it only superficially investigated.

The chapter is not intended to be an exhaustive review of every empirical study related to the role of television in children's attitude formation. Indeed, we have been quite selective. First, we have limited attention to studies which directly address the relationship between televiewing and social attitudes. At minimum, this required that the research contain a measure of television exposure or some kind of experimental manipulation of exposure, a measure of social attitudes or values, and an analysis of the relationship between the two. Second, we concentrated on studies in which the "effects", or dependent variables, manifested a strong affective or evaluative dimension. Thus, for example, children's feelings about a certain career – whether it would be appropriate for a girl to aspire to, or interesting, or fun – would be included. However, knowledge about that career, the skills required, the salary to be expected, would not be included. Measures of how much children like certain political candidates would be in; awareness of the issues in the campaign would be out. This is not to say that all evaluative measures were automatically included in our review, however. We concentrated on relatively broad, underlying, unifying social attitudes, while tending to ignore particularistic attitude objects. In the area of consumer socialisation, for instance, we do not look at the ability of television to increase the attractiveness of specific products, great though this ability may be. Rather, we consider the evidence relating viewing to perceptions of the truthfulness of advertising or to the acquisition of materialistic values.

We believe that values and attitudes are important, that they have the potential to influence the way children function in society, and that once established, they can carry over into adulthood. These are assumptions, however. Relatively little is actually known about either the behavioral implications or the long-term durability of any of television's effects on children, and this is particularly true of the sort of attitudinal effect with which we are concerned here. Towards the end of the chapter we will engage in some speculations about such long-term effects.

Our overall goals in this paper can be summarised as: (1) to present a useful analysis of some of the relevant research on how television influences children's social attitudes; (2) to specify and discuss some of the intervening variables and conditions which may mediate attitudinal effects of the medium; (3) to suggest some questions that need to be asked and the kinds of methods needed to answer them.

Research on television's attitudinal effects

EARLY WORK

The first, and perhaps still the most ambitious, investigation of the relationship between children's attitudes and their exposure to mass media was the series of reports known as the Payne Fund Studies (Charters, 1933), a nearly fifty-year-old collection of experimental, quasi-experimental, and survey research on the effects of motion pictures on children. Of the various Payne Fund volumes, the most germane here is the one by Peterson and Thurstone (1933), who looked for changes in attitudes toward various social objects as a result of viewing one or more of thirteen different films which the authors felt could be expected to influence different attitudes. For example, *Birth of a Nation* was expected to influence racial attitudes; *All Quiet on the Western Front*, attitudes toward Germans, and so on. Other social objects targeted in the study included capital punishment, war, crime, prohibition, and Chinese people. Four thousand junior and senior high school students participated in the study, which used a before–after design with no unexposed control group.

Among the findings and conclusions: (*a*) certain films (about half of those used) changed attitudes with but a single exposure (e.g. one showing of *Birth of a Nation* significantly decreased favorability toward blacks); (*b*) the attitudinal effects were cumulative (i.e. seeing two or three different movies with a more or less consistent position on a given topic caused more attitude shift than seeing one film); (*c*) although effects did decay, many endured for some time (after *Birth of a Nation*, for example, racial attitudes had still not returned to baseline levels eight months later).

Of course, Peterson and Thurstone's (1933) findings have been criticised. For example, questions have been raised about the use of quasi-experimental designs. But the realities of conducting research in the "real" world (as opposed to the laboratory) have convinced many in the field of communication research that we must learn to live with the weaknesses of such designs in order to gain from their strengths. As we

will see, a number of more current studies of television's attitudinal
effects have also used quasi-experimental designs. Similarly, one can ask
about the relevance of studies based on a single exposure to feature films
to a concern with television effects, since the latter are often conceived as
developing over many years of viewing. Still, there are movies as well as
other "single" programs on television which may be powerful enough to
influence children's attitudes and cognitions in a single exposure.
Indeed, both the existence and the resilience of single-program television
effects under "natural" (home) viewing conditions have been
demonstrated in recent years (e.g. Alper and Leidy, 1970). Finally,
regardless of any weakness in this early work, a look at the Payne Fund
studies on motion pictures and youth is worthwhile. Most of us can
probably benefit from the sense of *déjà vu* (or perhaps *déjà lu*) that such a
look will bring – a sobering realisation that we are still asking many of the
same questions raised by the Payne Fund researchers a half century ago.

Turning to the more recent literature on television's influence on
children's attitudes, there are several ways to organise this material, each
as artificial as the next. We have chosen to arrange it by subject matter
rather than, for example, age group or methodology, because the "problem orientation" of the field (Roberts and Bachen, 1981) has led the bulk
of the research to be indexed in this way. Within each subject area, we
will consider both methodological and age differences in the research.

POLITICAL ATTITUDES

In recent years a number of survey studies have dealt with the role of
television in political socialisation, although relatively few have directly
addressed the relationship between television viewing and political
attitudes or values. However, there is evidence of association between
television-viewing and political attitudes in some contexts for some
children.

Atkin (1977) found significant correlations (ranging from +0.26 to
+0.39) between third- to sixth-graders' self-reports of viewing political
commercials for US Presidential candidates on the one hand, and how
much they liked the candidates on the other. This result, he suggests, is
more likely to reflect an effect of commercials on attitudes than an effect
of prior attitudes on message selection, because children this young have
so few established predispositions to begin with. The Atkin study failed
to find a relationship between news-viewing and candidate popularity.
Conway, Stevens and Smith (1974), however, did obtain a positive
relationship between news-viewing and a related variable, the extent to
which children aged between nine and eleven stated a preference for a
specific political party rather than saying they were undecided or didn't
know what a party was. The relationship was limited to those with
greater than average interest in politics and did not emerge with nonnews television viewing. Such "partisanising" of children by television
may also be limited to younger children, since Chaffee, Jackson-Beeck,
Durall and Wilson (1977) failed to find it in teenagers.

A few studies have looked at attitudes toward government in general. Roberts, Hawkins and Pingree (1975) measured children's attitudes toward political conflict, rating the degree to which children accepted inter-party conflict as a legitimate part of the political process. Teenagers who reported more news and public affairs television viewing were more accepting of such political conflict. However, the relationship did not emerge among the younger children in the sample. Similarly, age again appears as a condition for effects in Rubin's (1978) survey of nine-, thirteen-, and seventeen-year-olds. The study included measures of attitudes toward the American president and the American Government, and measures of political efficacy and political cynicism. Controlling for age, Rubin found no significant correlations between those variables and either public-affairs viewing or general television viewing, with two exceptions: (1) there was a negative relationship (-0.17) between general viewing and cynicism among seventeen-year-olds, and (2) there was a positive relationship ($+0.21$) between public affairs viewing and attitudes toward the government among nine-year-olds. Finally, Atkin and Gantz (1978) report moderate correlations (none greater than $+0.20$) between television news viewing and interest in political affairs among children aged between six and eleven. The authors take these findings as evidence of a viewing effect, but admit the possibility of reverse or mutual causality.

Correlational evidence of this kind does not, of course, demonstrate causality. Moreover, we find no experimental evidence bearing on the question of television's impact on political attitudes. However, Alper and Leidy's (1970) quasi-experimental study of the effects of CBS television's "National Citizenship Test" is relevant here. They interviewed a large sample of high-schoolers before the airing of the program, which dealt with principles of United States constitutional law. Subsequent to the broadcast, they interviewed the same teenagers again and compared the responses of those who had seen the show with those who had not (self-selected groups). Their results showed not only significant increases in knowledge of the principles treated in the show (and no impact for principles not treated), but also an increased level of agreement with those principles. That is, attitudes toward the principles guaranteed by the Bill of Rights were more favorable. Further, these effects had some staying power. Although they did dissipate to some extent, they were still evident six months later. Although the Alper and Leidy study was not a true experiment, it certainly points in the direction of a causal influence from television on adolescents' political attitudes.

SEX-ROLE ATTITUDES

In recent years, considerable attention has been paid to the role of television in children's acquisition of sex-role related values and attitudes. The studies in this area have two basic thrusts: (1) to document effects of current television fare which, it is claimed, offers a biased and stereotyped view of women; (2) to demonstrate the ability of "alternative" programing approaches to cultivate a more egalitarian set of values.

There are signs that heavy viewing of American television content is associated with certain patterns of sex-role-related values. For instance, Beuf (1974) interviewed three- to six-year-olds about their career aspirations, and found that those who watched the most television were the most likely to prefer an occupation which conformed to common sex stereotypes (as defined by the author). This same general pattern is reported for children aged six to eleven by Freuh and McGhee (1975), who related amount of televiewing to responses on the IT scale, a measure of traditional sex-role adoption. They found that heavy viewers (25 hours or more per week) gave more "traditional" responses than did light viewers (10 hours or less).

The strongest evidence that television viewing influences children's sex-role attitudes comes from a longitudinal study of American adolescents (Morgan, 1980). Measures of television exposure (hours viewed in the "average day"), acceptance of sex-role stereotypes, and educational/ occupational aspirations, were taken over the course of two years, and the method of cross-legged panel correlations was applied to the data. The results support the view that television inculcates certain sex-role views, although the effects are limited to girls. Television viewing in the first year of the study significantly mediated girls' third-year attitudes – heavy viewers were more likely than light viewers to agree that men have more ambition than women, that women are happiest raising children, and so on. There was also, again only for girls, a relationship between first-year viewing and educational/occupational aspirations. Interestingly, however, the heavier viewers were the ones who two years later set their sights *higher*. This result, although predicted by Morgan on the basis of television's over-representation of professional women, still seems somewhat contradictory given the rather conservative influence generally demonstrated in both this and other studies. Indeed, Morgan concludes that, overall, his data favor the hypothesis that the "traditional" view of woman is what is being cultivated. In this case, it would seem, the hypothesis needs further specification in order to take gender differences into account.

The majority of the studies bearing on the other side of the question – the power of nonstereotypical television portrayals to move attitudes in non-traditional directions – are experimental in nature. The notable exception is Miller and Reeves' (1976) survey of grammar school children. They began with the fact that content analyses of US television programs show that men and women are portrayed differently on a number of social dimensions (e.g. women are less likely than men to be employed, more likely to be married, more likely to hold jobs which are less varied and have lower status, etc.) and then speculated about the possible effects of programs that counter this trend. They located five programs which featured women in traditionally male occupations, and measured exposure specifically to these shows. Their results revealed a strong positive association between frequency of viewing these counter-stereotypical programs and children's sense that it was "OK" for girls to aspire to the kinds of roles portrayed in the shows – school principal, police officer, park ranger, and television producer. Moreover, gender did not make a difference in exposure effects; boys were as accepting of nontraditional aspirations as were girls.

Experimental studies of television and sex-role attitudes have employed a range of stimuli, age groups, and dependent variables. Tan (1979) showed high school girls a series of 15 commercials emphasising the importance of physical beauty and then compared these girls' responses on measures of the perceived importance of physical attractiveness in attaining success in various contexts to the responses of girls who had not seen the commercials. Effects emerged for two of the measures: those who had seen the "beauty" commercials were more likely than unexposed girls to agree (*a*) that beauty was "personally desirable for me" and (*b*) that beauty was important "to be popular with men". Viewing the commercials did not affect girls' expectations concerning success as wife or success in a career. Using younger children (aged between eight and thirteen) of both sexes, Pingree (1978) looked at the effects of "nonsexist" commercials (e.g. showing women as athletes or doctors) on attitudes about women's roles and characteristics. The results of the study are complicated; they do show that nontraditional portrayals of women in commercials can alter sex role perceptions, though the effect in this study seemed to depend on the mental set (conveyed via instructions) under which the children viewed the ads, as well as their age and sex. The influence of commercials in this area has been demonstrated among even younger children by Atkin and Miller (1975), who showed children aged seven and ten a videotape of a children's show which contined commercials portraying a woman as either a judge, a computer programer, or a television technician. The dependent variable was the extent to which children felt women could attain the three jobs portrayed. When compared to children who saw no commercials, only the "judge" advertisement produced any effect. Seeing that commercial made children, especially older children and girls, more likely to say that a woman could be a judge.

There are also some experimental data on the influence of longer programs. Davidson, Yasuna and Tower (1979) showed a group of five-year-old girls one of three cartoon shows. One showed a girl in a reverse-stereotyped way, i.e. performing well at various traditionally male pursuits (sports, building a clubhouse); another showed a woman in stereotyped fashion; the third paid no particular attention to sex roles. The reverse-stereotyped version produced significantly less sex-stereotyping of personality characteristics than the other two programs, although as the authors admit, it is impossible to know for sure which of many differences, both related and unrelated to sex role perceptions, among the programs were responsible for producing the effects.

The recent extensive evaluation of *Freestyle*, a TV series produced by US public television with the express purpose of changing sex-role attitudes and perceptions, adds to the picture (Johnston, Ettema and Davidson, 1980). Nine- to twelve-year-old children participated in one of two main exposure conditions: in-school viewing only, and in-school viewing plus class discussion (home-viewing was not assessed). The viewing-only condition produced several attitudinal effects. Compared to unexposed children, viewers manifested increased acceptance of: boys in "helping roles", such as helping with housework or child care (effects on both sexes); girls performing mechanical tasks (both sexes); girls in the role of leader (effect on boys only); the idea of having more men in "female" jobs (girls only); husbands doing more household tasks (girls only). However, there were several targeted areas that were *not* changed, including attitudes toward girls in athletics, girls taking physical risks, more women in "male" jobs, wives doing more "male" household tasks, and the idea that wives as well as husbands should provide the financial support for the family. Not surprisingly, the effects of *Freestyle* in the viewing plus discussion condition were somewhat larger, and they extended to a wider range of variables.

RACIAL AND ETHNIC ATTITUDES

A few studies have directly examined relationships between television viewing and attitudes related to race, ethnicity, or nationality. Greenberg (1972) reports a survey that failed to find any association between viewing thirteen television programs with primarily black characters and nine- and ten-year-old white children's attitudes toward blacks. This lack of correlation held regardless of how much direct personal contact with blacks the children reported, a finding counter to Greenberg's expectation that television would have more influence among those with little direct experience.

As often happens, experimental evidence tends to contradict this no-effect result, indicating that there can be influences if one only has sufficient control over program production and distribution. One Canadian study of three- to five-year-olds (Gorn, Goldberg and Kanungo, 1976) placed inserts designed to create more favorable attitudes toward French Canadians and nonwhites (Orientals and Indians) into *Sesame Street* programs. After viewing, the children were shown a series of photographs of children from the inserts, and were asked to select those with whom they would most like to play. As predicted, the children who had seen the inserts were more likely to select either nonwhite or French Canadians (depending on the inserts) as playmates than were control-group children. The authors' interpretation of this finding was that the inserts changed racial attitudes. This may be the correct interpretation, but the conclusion would have been stronger if the photographs had shown unportrayed characters instead of children who appeared in the inserts, any of whose personal characteristics, including the simple fact that they had appeared on television, might have increased their attractiveness as playmates.

Other studies confirm the ability of educational programs aimed at improving racial attitudes to accomplish their purpose. Bogatz and Ball (1971) showed that two years of viewing *Sesame Street* led both black and white preschool children to more positive attitudes toward blacks and Hispanics. Mayes, Henderson, Seidman and Steiner (1975) found that exposure over an 8-week period to 16 half-hour episodes of a television series which purposely portrayed various minorities positively, increased six- to ten-year-olds' (again both black and white) acceptance of and friendliness toward those minorities. It seems clear that purposive programing can positively alter children's racial attitudes, especially with prolonged exposure (Graves, 1980). Furthermore, it appears that this sort of effort can influence attitudes toward people from foreign lands as well. Roberts and his students (1974), conducted a quasi-experimental evaluation of the children's series *The Big Blue Marble*, one purpose of which was to alter United States children's views of children from other cultures and parts of the world. A before–after (no control group) design was used, participants in the study being nine- to eleven-year-old American children. Overall, the postviewing measures showed (1) dramatic increases in estimates of the wellbeing of children from other parts of the world (they were seen as happier, better off, and so on); (2) decreases in various measures of ethnocentrism (e.g. after-viewing children were less likely to agree that American children were more fun, more interesting, more intelligent, and so on, than children from other countries); (3) younger children were more influenced than were older children.

Several studies have examined the effects of commercial programs on racial attitudes. Graves (1975) found that the attitudes of six- to eight-year-old white children were affected by a single exposure to commercial cartoons. Not surprisingly, positive portrayals produced positive attitude change and negative portrayals produced negative change. The reactions of these white children were more positive when blacks were in the minority than when they made up the majority of the cast of characters. The pattern of results was quite similar for black children: one exposure produced attitude change and the "integrated" setting produced more favorable attitudes than the all black (or nearly all black) setting. Leifer, Graves and Phelps (1976) report data on the effects of black portrayals in prime-time programs which confirm Graves' (1975) finding that the nature of the portrayal (i.e. whether blacks were portrayed in a positive or negative light) predicted the nature of the viewers' reactions. However (Leifer *et al.*, 1976) also found an effect for race that directly contradicted Graves' results (1975) – here black children were more likely to change in a negative direction than were white children. Graves (1980) notes this contradiction in a recent review, but points out the difficulty in directly comparing the two studies, because of various procedural differences, including subject selection, assignment to condition, and the nature of the stimuli used. These are, incidentally, but a few examples of the many barriers to comparability and synthesis that plague the children-and-television research literature in general.

MISCELLANEOUS STUDIES

There remains a miscellany of studies which do not categorise as well by subject as the ones we have reported, but which, nonetheless, do bear mention before we move on. First, concerning occupation and career-related attitudes, DeFleur and DeFleur (1967) interviewed six- to thirteen-year-olds in order to ascertain both their knowledge of the role of certain occupations and, more to the point, their perceptions of the relative status of those occupations. They did not measure television exposure, but instead compared the accuracy of children's notions about three sets of occupations: those with which children might well have had some personal contact (e.g. minister, teacher, grocer), those for which the only likely source of contact was television (e.g. judge, reporter, headwaiter), and those occupations which are widely understood by adults and which are part of the general culture but are rarely seen in action by children. Children's status rankings of the "television" occupations were much closer to those of parents and experts than with either the "personal contact" or "general culture" occupations, leading the authors to suggest that television is a "more potent source" of occupational status knowledge than the other two. Much of the research described earlier in the section on sex-role attitudes is also relevant to the world of work (e.g. Atkin and Miller, 1975; Johnston et al., 1980; Morgan, 1980) and indicates that television may influence occupational attitudes. At any rate, the argument is made that television does provide occupational information, however stereotypical or inaccurate, and that the distorted images it provides may well persist into adulthood (Leifer and Lesser, 1976).

Of the many studies concerned with the effects of TV advertising on children (cf. Adler, Friedlander, Lesser, Merringoff, Robertson, Rossiter and Ward, 1977; Comstock et al., 1978; Roberts and Bachen, 1981), only a few have directly considered the impact of commercials on long-term consumer values and attitudes. There are data suggesting that higher levels of viewing among grade school children are associated with more favorable attitudes toward commercials and greater trust in them (Atkin, 1975a; Rossiter and Robertson, 1974), though of course causation is in doubt on the basis of these survey data. Experimental evidence shows that levels of trust in commercials can be lowered by brief television announcements (Christenson, 1980a), but this study did not really address the issue of cumulative impact. In another experiment, Goldberg and Gorn (1977) dealt with a possible unintended adverse effect of advertising. They report that exposure to toy advertisements increased preschoolers' tendency to choose potential friends on the basis of what toys they possessed, even when a price had to be paid in terms of how pleasant a person the potential friend was. However, we don't know if such "materialism" results from patterns of viewing, or what its implication for adult behavior might be.

Some attention has been paid to the effects of commercials on health-related attitudes. One survey of eight-, ten-, and twelve-year-olds revealed small but significant correlations, controlling for age, between exposure to over-the-counter drug commercials (extrapolated from an hours-per-day television viewing measure) and such health attitudes as illness anxiety, belief in the efficacy of medicine, and attitudes toward taking medicine (Robertson, Rossiter and Gleason, 1979). This supports the earlier work of Atkin (1975b), who estimated exposure to drug commercials by multiplying overall viewing by a measure of attention to sample commercials, and found for children aged between nine and eleven that exposure correlated with illness anxiety, approval of medicine, and belief in the efficacy of medicine. Again the correlations were small (none higher than +0.14), but they did survive several simultaneous controls. However, as with many other attitudinal relationships, age may be a factor here, since a three-year panel survey found no such relationships among teenage boys (Milavsky, Pikowsky and Stipp, 1975).

Finally, Bailyn (1959) found a relationship, among boys only, between the use of pictorial media, including television, movies and comic books, and the tendency to see people in "black and white" terms. Boys high in media use were more likely to agree with statements like "People are either all good or all bad", or "There are only two types of people in the world, the weak and the strong". The relationship with media use was stronger for boys who reported more than an average number of personal problems and for those who were relatively extrapunitive, suggesting two additional variables which may specify television's attitudinal effects on children.

Conditions and intervening variables

The assumption of most of the studies summarised here is that there ought to be some match between the content of television and the attitudes and values of young viewers. This is an obvious notion, of course, and by itself not very interesting. The basic effect has been demonstrated often enough; we can safely say that television frequently has an influence. The interesting and difficult questions, however, involve the conditions under which one is most likely to find it.

THE DEFINITION AND MEASUREMENT OF THE TELEVISION STIMULUS

One of the barriers to making useful generalisations from survey studies is the wide disparity in measures of television exposure. Exposure has been measured in numerous ways. There are daily television logs, estimates of hours usually watched per day, hours watched yesterday, reports of favorite programs, reports of viewing specific types of programs (e.g. news or violent shows), and more. To an extent, the choice of measure depends on the research question. If the interest is in attitudes toward public figures, then it is more logical to measure exposure to public affairs and news programs than overall television viewing. There is little reason to expect situation comedies and detective dramas to influence these sorts of attitudes.

As a general rule, the more specific one's measure of exposure, the more likely it is that whatever relationships there are will be reflected in the data. This is demonstrated quite neatly by Comstock and his colleagues (1978) in their review of the survey research on violence viewing and aggression in children. The measures of exposure used in these studies varied from the global presence or absence of television in the home, through self-reports of overall viewing, to more direct measures of actual violence viewing such as the number of programs viewed in the last week from a list of the 20 most violent ones available on television. It turned out that the latter types of measure, the more direct, were far more likely to reveal correlations between viewing and aggressiveness. As an example of this principle in the literature surveyed in the preceding pages, Conway et al. (1975) found that news-viewing predicted political partisanship, but that nonnews viewing did not.

Of course, the Cultural Indicators group (Gerbner and Gross, 1976; Gerbner and his co-authors (1978, 1979, 1980) have typically used overall hours per day as the measure of exposure. They have looked for broad correlations between people's normal, everyday television use, assumed to be nonspecific, nonselective, habitual behavior, and certain systematic pictures of the world as seen on television. This approach makes sense if their assumptions about the nature of viewing holding true, and if there really are coherent television "biases". However, even if most viewing is random and unselective, much is not. One can never be sure that people are really watching a true cross-section of "typical" television fare, and hence there is no certainty that viewers are seeing the biased picture of the world the effects of which are being investigated. Broad measures of overall viewing tend to overlook and underestimate effects; even more important, they can never tell us *which* content is having *which* effects, a very important issue indeed for those concerned with how to prevent adverse and/or promote positive television influences.

If one is looking for isomorphism between long-term television viewing and certain biases or pictures of the world inherent in "typical" content, one is much more likely to find it if the television "world" is coherent and unified. Such a world is much more likely to occur in television systems whose offerings are narrow, whose programs and channels tend to be very similar in content and point of view. This was true to some extent in the USA during the 1960s; there were three dominant television networks whose daily offerings were indeed very similar in general approach and treatment. Today, however, in addition to a number of local stations, it is not unusual to find cable systems which offer twenty or thirty channels, including all-sports, all-news, all-weather, all-movie, religious, educational, financial, and so on. And in many homes one can find an assortment of video-games absorbing children's "television time". These new options are certainly changing patterns of viewing, and their variety will make it more and more difficult to define "typical content", much less identify its "mainstream". In this kind of environment, gross measures of viewing cannot be expected to correlate very highly with effects, attitudinal or otherwise.

SOME AGE-RELATED INTERVENING VARIABLES AND CONDITIONS

Relatively few of the studies reviewed here report results separately for different age groups, and those that do present no clear picture of the relationship between age and amount of television influence. There is evidence of a positive relationship between age and effect (Atkin and Miller, 1975, Roberts et al., 1975), of a negative relationship (Roberts et al., 1974), of little relationship (Atkin and Gantz, 1978), and evidence that the relationship depends on the measures involved (Rubin, 1978). In fact, there is probably no simple relationship between age and the nature or extent of television's attitudinal effects. Many substantive changes occur with increasing age, some of which act to increase influenceability, some of which decrease it. The overall effect of age in this context is bound to be part of a "compromise" among these distinct trends, and the nature of that compromise will depend on various other factors. For now, though, we will be content to mention a few of the age-related trends which may mediate television effects on attitudes.

Perhaps the most obvious age-related trend is the increase in comprehension of television content. The developmental improvement in the ability to comprehend audio-visual narratives is as well-documented as it is unsurprising (see Comstock et al., 1978, for a review), but it is not entirely clear how comprehension may relate to television's affective impact. First, comprehension is a very broad concept. Collins (1979) describes "mature" comprehension as consisting of three conceptually distinct abilities: (a) selection of the essential information from a program; (b) ordering the scenes in the program according to some organising scheme; (c) making inferences that go beyond what is concretely presented in order to unify the discrete elements into a meaningful whole. It seems clear that these sorts of abilities will tend to facilitate the effects of children's programs which purposely try to alter attitudes and values, such as the Freestyle series on sex-roles (Johnston et al., 1980). In these cases, the value-altering content tends to be central to the program, and explicit arguments are made in support of the advocated point of view. The more fully the child understands the program, other things being equal, the greater should be its impact.

In the case of long-term entertainment viewing, however, comprehension may not play such a straightforward role. Here the effects on values come from two sources: (*a*) from "central" events and content, as when a story contains a moral lesson which can be abstracted from the plot; and (*b*) from a variety of incidental, plot-irrelevant details, comments, minor portrayals, and so on. Recent research indicates that the ability to abstract moral lessons embedded in prime-time drama is dramatically a function of age, generally not appearing at all before the age of seven or eight years (Christenson, 1980b), which implies that this kind of attitudinal effect would probably increase with age. Conversely, it has been suggested that the influence of "incidental" material may actually decrease with age, since it is this sort of information that older children can ignore but younger ones cannot (Hawkins and Pingree, 1981). This is an interesting hypothesis that, to our knowledge, remains untested in the attitudinal realm.

Roberts, (1982; Roberts, Bachen and Christenson, 1978) has hypothesised still another age-related change in how children process information which could mediate the attitudinal influence of television content. One of the initial steps in adult processing of a message, he argues, is the making of an attribution about the intent of that message, that is, deciding whether it is primarily intended to inform, educate, entertain, or persuade. The nature of subsequent processing, it is contended, will be influenced by that attribution. For example, one person might see a particular program as attempting only to provide light entertainment, while another perceives the same program as an attempt to persuade viewers to adopt a particular position on some social issue. We would expect these two viewers to engage in very different processing of and responses to program content. Young children, however, are hypothesised not to make such differentiations among message types, Rather, the ability to distinguish between different intents appears to be ontogenetic (cf. Roberts, 1982). Several studies, for example, demonstrate that the persuasive intent of commercials is seldom recognised by children below the age of seven. Rather, younger children appear to "understand" commercials in informational terms (e.g. "They tell us about things we can buy"). Although we know of no research which has addressed the issue, it also seems reasonable to expect relatively young children to process a good deal of entertainment content in primarily informational terms. If such is the case, then we would expect the attitudinal impact of much entertainment programing to be inversely related to age.

This hypothesis dovetails with similar predictions concerning the role of perceived reality and the development of critical evaluation skills. For example, a number of studies have hypothesised that the more real a particular television show is perceived to be, the more likely it is to influence the learning and/or behaviour of children, and there are data indicating that, under some circumstances, this is the case (for reviews see Comstock *et al.*, 1978; Hawkins, 1977). However, Hawkins and

Pingree (1982) also note that the results of studies that have attempted to manipulate perceived reality are rather complex, sometimes showing direct effects, sometimes finding that perceived reality interacts with such variables as involvement with the program and/or relevance of the portrayed information, and sometimes producing reverse effects, with materials perceived as having less reality having greater influence. These authors argue that what may be important in the various "perceived reality" studies is not the degree to which children perceive television to present the real world, but the degree of mental "activity" invested in such processes as evaluating and comparing income information with existing knowledge, searching for patterns or for confirming or disconfirming information, and so on. Such activity, it is posited, is a function of how "involved" with the program content the child may be, lower involvement implying less activity, thus a greater likelihood of television influence. Conversely, more "active" processing implies more critical processing, thus less influence.

The Hawkins and Pingree (1982) discussion is couched in terms of the child's construction of social reality, a construct which certainly includes social attitudes such as we have been discussing in this chapter. Moreover, the development of the kinds of "critical consumer processes" they suggest is certainly congruent with the development of the attribution strategies Roberts (1982) hypothesises. Here, then, is an area in which a good deal of empirical research on mediators of the attitudinal effects of television could be conducted.

Another age-related factor that deserves attention has to do with the presence and strength of children's preexposure attitudes and values. It has been suggested that the traditional premise of the "law of minimal effects", which states that mass media act mainly to reinforce existing predispositions (Klapper, 1960), may not be very relevant to young persons, simply because they have so few predispositions (Chaffee, Ward and Tipton, 1970). We would add that the process of acquiring these predispositions is gradual and continuous throughout childhood. A child of five years will have fewer prior attitudes on fewer topics than will a ten-year-old, who in turn, is likely to have fewer than a fifteen-year-old. We also expect that what attitudes the young child has will be less firm, less integrated within a total cognitive map. For these reasons, then, attitude formation and change should be easier to accomplish among younger children. Of course, the notion of "preexposure attitudes" is a fiction, in a way, since television is introduced so early into children's lives and since its presence is so constant. Still, we can surmise that new content or further exposure at a given time will have more of an impact on younger children.

THE ROLE OF OTHER AGENTS OF SOCIALISATION

Television is obviously not the only teacher of values, attitudes and norms. It interacts and "competes" with a number of other socialisation agents, including parents, extended family, peers, school, churches, other media and so on. We suspect that television is a sort of "default" mechanism in the system. Its influence will be substantial primarily when the issues involved do not receive much attention from primary socialisation agents. However, if a child's parents or peers take up an issue, they are likely to prevail. There are not many data on this point, but the findings of Tolley (1973) are relevant. He surveyed seven- to thirteen-year-olds, measuring their knowledge about and attitudes toward the Vietnam War, as well as the sources of the knowledge and attitudes. Television was very influential as a source of information about the war; those who watched television news regularly knew more facts. But television did not seem to have anything to do with support for the war. Rather, attitudes toward the war were determined by parents. Television supplied the information, but parents determined how the information was interpreted and applied to the morality of the war (Comstock *et al.*, 1978). A further suggestion that television takes on a secondary role when other important socialisation agents get involved comes from the finding that children are more likely to cite television as an important source of political attitudes and opinions when the specific topic is not likely to be salient in interpersonal communication (Connell, 1971).

We wish to emphasise, however, that to limit television to this sort of role is not to relegate it to unimportance. There are many topics about which television has the first or only word. Precisely what those topics might be will vary from child to child. Some parents don't talk about politics, others do; some do not talk about sex or sex-roles, others do; some don't discuss health matters, others do. Even if parents and schools do eventually deal with a particular attitude object, if television is first, its influence may be lasting and significant in that it may provide some of the standards against which subsequent information is judged. An example might be the formation of nutrition attitudes and influences which are addressed very early by television commercials. Some rather resilient attitudes and even habits may develop by the time parents and schools get into the picture to actually combat the sugary view of the commercials.

DIRECT PERSONAL EXPERIENCE

Though not an agent of socialisation in the same sense that parents or schools are, direct personal experience is obviously a strong influence on attitudes and values. Currently, we know little about the part that direct experience plays in mediating the effects of television on children. Hawkins and Pingree (1982) describe three notions concerning the role of direct experience: (*a*) that for any given television message some degree of *confirmation* by direct experience may be necessary, or effects will not occur; (*b*) that experience may provide *disconfirmation* of television's message, thereby acting to *prevent* effects; and (*c*) that television presents a *mainstream* of beliefs which most people share for reasons other than television (including direct experience), therefore, its effects will be felt only on those who diverge from that mainstream. These three processes are not mutually exclusive, of course. No doubt they each occur at times: the problem is predicting when.

It may be useful to introduce a distinction between the absolute effects and the relative effects of television. The influence of television relative to other agents of socialisation is likely to be greatest when those other agents are not concerned with the issue at hand. But this doesn't necessarily mean that the teachings of television have any great power to determine the way a child functions in society. The attitudes learned under this condition may simply be isolated and irrelevant to the child. Perhaps the greatest absolute impact in this sense, comes when the rest of the environment provides confirmation of the television message or when the television provides a stimulus for, say, family discussion, thereby focusing even greater attention on the television message. Finally, both absolute and relative impact should be low when experience and other sources directly disconfirm the television show.

FINAL COMMENTS ON INTERVENING VARIABLES AND CONDITIONS

The list of factors that have been suspected of mediating television's effects on children is extensive. We have discussed only a few here. In the studies on attitudes reviewed earlier, we encountered several of them: gender, especially in interaction with topic (Morgan, 1980; Atkin and Miller, 1975); race (Leifer *et al.*, 1976); the presence of personal problems and extrapunitiveness (Bailyn, 1959), the context in which a value is to be applied (Tan, 1979; Atkin and Miller, 1975); level of interest in the subject matter (Conway *et al.*, 1975); interpersonal discussion after viewing (Johnston *et al.*, 1980). Other likely candidates abound, even though they have not appeared here. If the list of mediating variables seem large, the interactions among them are legion. Obviously, it is impossible to investigate all of the possible interactions, and luckily it is not necessary since so many of the concepts overlap. We have discussed a few of the variables that we believe will turn out to be important mediators of attitudinal effects on children. The important thing now is to accept that simple answers to the questions we have been asking are not forthcoming, and to investigate some of the complexities.

Concluding comments and speculation

This chapter began with a characterisation of the research on children and television as based largely on a concern for children's well-being. For years this concern has found its outlet primarily in special-topic, policy-related studies, employing different measures, different conditions, children of different ages, and so on, depending on the policy need. The unfortunate result is that it is very difficult to synthesise our knowledge; there are so many differences between studies that they are virtually impossible to compare. We find our knowledge growing out, but not up. We need not only standardisation of the few good measures we have, but much more continuity in the research, much more of an effort to take prior, related work as a departure point for future studies.

One of the obstacles for this in the United States has been the dependence of researchers on government agencies for financial support. In general, the research has addressed the specific policy-making goals of the Federal Communications Commission, the US Surgeon General, or the Federal Trade Commission. Another result of such dependency is the concentration of research on issues that relate more to questions concerning possible changes in the content and structure of the medium than to interventions that might change the way children process the messages of the medium. As a practical matter, if one is truly concerned about the effects of televison on children, it may prove more feasible to improve children's information processing skills than it is to effect any broad, socially significant changes in the television diet that is offered. Indeed, efforts along these lines have recently been made (see, for example, *Journal of Communication*, Summer, 1980). At any rate, it is apparent that our limited understanding will only be remedied if more attention is given to mediating factors, and less to the simple demonstration of effects (Chaffee, 1976; Withey and Abeles, 1980).

As Withey (1980) has observed, there is probably no such thing as a "television message". People perceive the same material in different ways. With respect to our current focus on social attitudes and values, it is inevitable that different children will have different perceptions of the positiveness of a certain portrayal or will derive different value lessons from a program. The attitudinal effect of a given program or of a certain steady diet will vary according to the relationship between the "television view" and the view given by the various other agents that supply information to the child. If the subject is, for instance, the desirability of women having a certain career, a lukewarm portrayal or statement on television about that issue may affect those who are positively disposed in a negative direction and those who are negatively disposed in a positive direction. It also is possible that groups who feel that they are negatively portrayed on television are presented in an even more negative way to many children by the rest of their social environment. In such a case, the effects of the television portrayals would, theoretically, be positive. If we are to ever be able to say what the effects are, we clearly need better measures of the nontelevision influences on children's attitudes and values. This includes data on parents, peers, schools, church, and any other relevant elements of the child's ongoing social milieu. Future research, both survey and experimental, should make an effort to gather these sorts of data.

Finally, it has become almost mandatory for reviews of the literature on children and television to call for more longitudinal studies, and this review is no exception. For most people, television viewing is an almost daily activity, Hence it is likely that many attitudinal effects of viewing are cumulative. To the extent that the view of the world presented by television programing influences the formation of children's social attitudes, much of that influence probably occurs over weeks, months, years of exposure. We cannot expect to understand television's role in attitude formation fully until we begin to conduct studies that include time as a variable.

References

Adler, R. P., Friedlander, B. Z., Lesser, G. S., Merringoff, L., Robertson, T. S., Rossiter, J. R. and Ward, S. (1977). *Research on the Effects of Television Advertising on Children*. US Government Printing Office, Washington, D.C.

Alper, W. S. and Leidy, T. R. (1970). The impact of information transmission through television. *Public Opinion Quarterly* **33**, 556–62

Atkin, C. K. (1975a). "Effects of television advertising on children – survey of children's and mother's responses to television commercials". Technical Report No. 8, Michigan State University

Atkin, C. K. (1975b). "The effects of television advertising on children: survey of Pre-adolescents' responses to television commercials". Technical Report, Michigan State University

Atkin, C. K. (1977). Effects of campaign advertising and newscasts on children. *Journalism Quarterly* **54**, 503–08

Atkin, C. K. and Gantz, W. (1978). Television news and political socialization. *Public Opinion Quarterly* **42**, 183–98

Atkin, C. K. and Miller, M. (1975). "The effects of television advertising on children: Experimental evidence". Paper presented at Ann. Meet. Int. Commun. Assoc., Chicago, Ill.

Bailyn, L. (1959). Mass media and children: A study of exposure habits and cognitive effects. *Psychological Monographs* **73** (1. whole No. 471)

Beuf, A. (1974). Doctor, lawyer, household drudge. *Journal of Communication* **24**, 142–45

Bogatz, G. A. and Ball, S. J. (1971). *The second year of Sesame Street: A continuing evaluation.* 2 Vols. Educational Testing Service, Princeton, N.J.

Butler, M. and Paisley, W. (1980). *Women and the mass media: Sourcebook for Research and Action*. Human Sciences Press, New York

Chaffee, S. H. (1976). Comparing television to other agencies of socialization. Unpublished manuscript, Mass Communication Research Center, University of Wisconsin

Chaffee, S. H., Ward, L. S. and Tipton, L. P. (1970). Mass communication and political socialization. *Journalism Quarterly* **47**, 467–59

Chaffee, S. H., Jackson-Beeck, M., Durall, J. and Wilson, D. (1977). Mass communication in political socialization. *In* S. A. Renshon (ed.) *Handbook of Political Socialization: Theory and Research*, pp. 223–58. Free Press, New York

Charters, W. W. (1933). *Motion Pictures and Youth: A Summary*. Macmillan, New York

Christenson, P. G. (1980a). The effects of consumer information announcements on children's perceptions of commercials and products. Unpublished doctoral dissertation, Stanford University

Christenson, P. G. (1980b). Prosocial themes in prime time: Are kids getting the message? Unpublished manuscript, Department of Speech Communication, The Pennsylvania State University

Collins, W. A. (1979). Children's comprehension of television content. *In* E. Wartella (ed.) *Children Communicating: Media and Development of Thought, Speech, Understanding*, pp. 21–52. Sage, Beverly Hills

Comstock, G., Chaffee, S., Katzman, N., McCombs, M. and Roberts, D. (1978). *Television and Human Behavior*. Columbia University Press, New York

Connell, R. W. (1971). *The Child's Construction of Politics*. University of Melbourne Press, Melbourne

Conway, M. M., Stevens, A. J. and Smith, R. G. (1975). The relationship between media use and children's civil awareness. *Journalism Quarterly* **52**, 531–38

Davidson, E. S., Yasuna, A. and Tower, A. (1979) The effects of television cartoons on sex-role stereotyping in young girls. *Child Development* **50**, 597–600

DeFleur, M. L. and DeFleur, L. B. (1967). The relative contribution of television as a learning source for children's occupational knowledge. *American Sociological Review* **32**, 777–89

Freuh, T. and McGhee, P. E. (1975). Traditional sex role development and amount of time spent watching television. *Developmental Psychology* **11**, 109

Gerbner, G. and Gross, L. (1976). Living with television: The violence profile. *Journal of Communication* **26**(2), 172–99

Gerbner, G., Gross, L., Jackson-Beeck, M., Jeffries-Fox, S. and Signorelli, N. (1978). Cultural indicators: Violence profile No. 9. *Journal of Communication* **28**(3), 176–207

Gerbner, G., Gross, L., Signorelli, N., Morgan, M. and Jackson-Beeck, M. (1979). The demonstration of power: Violence profile No. 10. *Journal of Communication* **29**(3), 177–96

Gerbner, G., Gross, L., Morgan, M. and Signorelli, N. (1980). The "mainstreaming" of America: Violence profile No. 11. *Journal of Communication* **30**(3), 10–29

Goldberg, M. E. and Gorn, G. (1977). "Material vs. social preferences, parent-child relations and the child's emotional responses: Three dimensions of responses to children's TV advertising". Paper presented at 5th Ann. Telecomm. Policy Res. Conf., Airlie House, Va.

Gorn, G. J., Goldberg, M. E. and Kanungo, R. N. (1976). The role of educational television in changing the intergroup attitudes of children. *Child Development* **47**, 277–80

Graves, S. B. (1975). Racial diversity in children's television: Its impact on racial attitudes and stated program preferences. Unpublished doctoral dissertation, Harvard University

Graves, S. B. (1980). Psychological effects of black portrayals on television. *In* S. B. Withey and R. P. Abeles (eds) *Television and Social Behavior: Beyond Violence and Television*, pp. 259–89. Lawrence Erlbaum, Hillsdale, N.J.

Greenberg, B. S. (1972). Children's reactions to television blacks. *Journalism Quarterly* **49**, 5–14

Hawkins, R. P. (1977). The dimensional structure of children's perceptions of television reality. *Communication Research* **4**, 299–320

Hawkins, R. P. and Pingree, S. (1982). TV influence on social reality and conceptions of the world. Mass Communication Research Center, University of Wisconsin. Chapter prepared for *Television and behavior: Ten Years of Scientific Progress and Implications for the 80's*. US Government Printing Office, National Institute of Mental Health. (In press)

Johnston, J., Ettema, J. and Davidson, T. (1980). *An evaluation of "Freestyle": A television series to reduce sex role stereotypes*. Inst. Soc. Res. University of Michigan, Ann Arbor

Klapper, J. T. (1960). *The Effects of Mass Communication*. Free Press, Glencoe, Ill.

Leifer, A. D. and Lesser, G. S. (1976). The development of career awareness in young children. *NIE Papers in Education and Work: No. 1*. National Institute of Education, Washington, D.C.

Leifer, A. D., Graves, S. B. and Phelps, E. (1976). Monthly project of critical evaluation of television project. Unpublished manuscript, Center for Research in Children and Television, Harvard University

Mays, L., Henderson, E. H., Seidman, S. K. and Steiner, V. S. (1975). An evaluation report on *Vegetable Soup*: The effects of a multiethnic children's television series on intergroup attitudes of children. Unpublished manuscript, New York State Department of Education

Milavsky, J., Pikowsky, B. and Stipp, H. (1975). Television drug advertising and proprietary and illicit drug use among teenage boys. *Public Opinion Quarterly* **39**, 457–81

Miller, M. M. and Reeves, B. (1976). Linking dramatic TV content to children's occupational sex-role stereotypes. *Journal of Broadcasting* **20**, 35–50

Morgan, M. (1980). Longitudinal patterns of televison use and adolescent role socialization. Unpublished doctoral dissertation. University of Pennsylvania

Peterson, R. C. and Thurstone, L. L. (1933). *Motion Pictures and the Social Attitudes of Children*. Macmillan, New York

Pierce, C. M. (1980). Social trace contaminants: Subtle indicators of racism in TV. *In* S. B. Withey and R. P. Abeles (eds) *Television and Social Behavior: Beyond Violence and Children*, pp. 249–57. Lawrence Erlbaum, Hillsdale, N.J.

Pingree, S. (1978). The effects of non-sexist television commercials and perceptions of reality on children's attitudes about women. *Psychology of Women Quarterly* **2**, 262–77

Roberts, D. F. (1974). Communication and children: A developmental approach. *In* I. de Sola Pool and W. Schramm (eds) *Handbook of Communication*, pp. 174–75. Rand McNally, Chicago

Roberts, D. F. (1982). Children and commercials: Issues, evidence, intervention. *Prevention in Human Services* (In press)

Roberts, D. F. and Bachen, C. (1981). Mass communication effects. *Annual Review of Psychology* **32**, 307–56

Roberts, D. F., Herold, C., Hornby, M., King, S., Stern, D., Whiteley, S. and Silverman, L. T. (1974). Earth's a Big Blue Marble: A report of the impact of a children's television series on children's opinions. Unpublished manuscript, Institute for Communication Research, Stanford University

Roberts, D. F., Hawkins, R. P. and Pingree, S. (1975). Do the mass media play a role in political socialization? *The Australian and New Zealand Journal of Sociology* **11**, 37–42

Roberts, D. F., Bachen, C. M. and Christenson, P. G. (1978). "Perceptions of and cognitions about television commercials and supplemental consumer information". Testimony prepared for the Federal Trade Commission Rulemaking Hearings on Television Advertising and Children. San Francisco, Calif.

Robertson, T. S., Rossiter, J. R. and Gleason, T. C. (1979). Televised medicine advertising and children. *Journal of Consumer Research* **6**(3), 247–55

Rossiter, J. R. and Robertson, T. S. (1974). Children's television commercials: Testing the defenses. *Journal of Communication* **24**(4), 137–44

Rubin, A. M. (1978). Child and adolescent television use and political socialization. *Journalism Quarterly* **55**, 125–29

Surgeon General's Scientific Advisory Committee. (1972). *Television and Growing Up: The Impact of Televised Violence*. US Government Printing Office, Washington, D.C.

Tan, A. S. (1979). TV beauty ads and role expectations of adolescent female viewers. *Journalism Quarterly* **56**, 283–88

Tolley, H. (1973). *Children and the war: Political socialization to international conflict*. Teachers College Press, New York

Withey, S. B. (1980). An aerial view of television and social behavior. *In* S. B. Withey and R. P. Abeles (eds) *Television and Social Behavior: Beyond Violence and Children*, pp. 291–301. Lawrence Erlbaum, Hillsdale, N.J.

Withey, S. B. and Abeles, R. P. (eds) (1980). *Television and Social Behavior: Beyond Violence and Children*. Lawrence Erlbaum, Hillsdale, N.J.

Social Learning from Everyday Television

Grant Noble

Rationale and overview

I am going to adopt a highly individualistic approach in this chapter which is based on my research and thinking during the past four years and continues the line of argument of my book *Children in Front of the Small Screen* (1975). I may delight some readers, but I expect I shall offend others: it has always been my view that the bland paper, like the bland hamburger, may have a semi-universal appeal but will eventually fail to please anyone. In my book I advanced the argument that television today replaced the extended kin of yesterday. Television portrayals allow the viewer contact with characters which represent the wider society.

I shall argue that all television teaches. I hold to this view notwithstanding the recent spate of polemic texts that appeal to well-heeled and puritanically neurotic people who worry about the sheer amount of time that children, in particular, spend viewing: *The Plug-in-Drug* (Winn, 1977) is a classic example of this viewpoint. Social science claims to build a cumulative picture of social events over research time, and recent polemics castigating television copy the negative stance adopted by the early researchers such as Maccoby (1951, 1954) who first studied the impact of television when it was first introduced in the United States. The polemicists have tended to ignore the slow but sure progress which has been made in the communication field since then. As Katz and Foulkes (1962) pointed out, the consequences of viewing so-called "escapist" television may well not coincide with the motive for watching. The most systematic studies concerned with the consequences of viewing have been conducted by those adhering to the "uses and gratifications" school. Despite the possible tautology – that is, the uses found for television are said to satisfy needs, and these in turn are derived from the uses which people find for television – their research strikes this writer as providing a more effective theoretical stance than any alternative approach.

Some definitions

Learning as a concept has been considerably used and abused, in part because of a lack of an understanding of the concept embodied in the word. Etymologically, the root of the word "learn" is identical to that of "teach". Both concepts are manifest in the word "lore", which is a body of facts or traditions as in folklore. It was my argument in 1975 that television is the prime socialisation agent of this day, a concept which Gerbner and his colleagues (Gerbner and Gross, 1976) have consistently reiterated. For better or worse television provides the folklore of today. Yet the dictionary also tells us that to learn or teach is literally "to lead someone on his way". From this are derived the two words "track" (or rut) caused by a wagon, derived from the Latin *lira* or "furrow", and the "last" which guides the shoemaker. In each case it can be argued that the concept embodied in the word "learn" involves guiding oneself along a specific path. The argument presented in this chapter will be that television today does guide viewers along the path of life. It gives some folklore which is representative of society's mores; it provides examples of appropriate behaviours for social settings beyond the viewer's direct experience, and it exposes viewers to people, places and events from the wider society. Such knowledge marks membership of the society. It is also worth noting that the word "delirium" comes from the same root as "learn", the prefix *de* simply meaning that the person has lost his path or way.

In this chapter therefore we shall trace a particular path of argument. First, working within a conventional definition of the word "learn", studies will be presented to show that children can acquire specific information from particular television programmes. Second, we shall broaden the argument to show that not only is specific information learned, but that moral messages from a televised programme are also acquired by the child audience. We hope to demonstrate that the latter kind of learning from television provides a way for children to acquire folklore and moral values.

Once this goal has been achieved we shall be in a position to explain why certain fictional programmes are better teachers than are well intentioned documentaries. Most researchers still seem to adopt the attitude that fictional representations cannot be other than escapist, while reality materials are not. (See, for example, Schramm, Lyle and Parker, 1961). My opinion is that many researchers ignore the unique qualities of television as a medium. I believe that television works best when it personalises an issue, as in the case of *Holocaust*, which provides the example used in the present chapter. Furthermore, television may work through feelings rather than through thoughts, these being but two of the primary ways of knowing outlined by Jung (1921) as being mutually antagonistic ways of apprehending reality.

When the argument has advanced thus far, a switch will be made towards the phenomenological approach of "uses and gratifications". We[1] shall report that children go to their television sets in order to learn, a motive at least as important as attempting to escape from everyday reality. What children say they learn will be reported in some detail. Finally I shall report on the way children use television characters to assist their moral development. This research will show how television lore is comparable to folklore in providing examples of moral behaviour which help the child to acquire a more mature morality.

Conventional learning from entertainment programming

Noble and Creighton (1981) have recently been involved with a series of nature programmes, *Australia Naturally*, which were produced for an Australian country station (Channel 9/8, Tamworth). The producer made a pilot programme which was evaluated by a child audience. Following modifications made in the light of comments about the pilot programme, a series of six further programmes have been aired. We asked 240 children from two schools and ranging in age from seven to eleven years to give us their opinions about the series; we were aiming to produce a second, and better, series of programmes. Throughout this action-orientated research we have had the ear of both television management and production staff.

Because the producer still sees his role as that of teacher, we were able to investigate whether or not the complex concepts such as symbiosis and adaptation which underpinned the series were learnt by the viewing children. First, we simply compared children who had seen the programmes with those who had not. We looked at their answers to factual questions concerning the content of a number of the programmes in the series. A sample of results from this research will suffice to make our point. Table 1 compares the answers given by viewers and nonviewers (comparable in all other respects), to the questions listed in the table. The children received a number of forced-choice and open-ended questions.

As the table shows, children who watched the programmes did acquire factual information about animals. The questions were selected to sample a variety of the broadcast programmes, which accounts for the variation in sample sizes. Questions were usually answered in schools on the day after transmission. Yet, as can be seen, children did not learn what the word "symbiosis" meant even though the word was laboured by the presenter. It appears that symbiosis is too difficult a concept for the seven- to eleven-year-olds sampled. For all the other questions, which concerned simple facts, viewers acquired the information from watching the shows.

---

[1] I hope the reader will forgive the "we". Most of the research projects reported herein were joint efforts.

Table 1 Comparison of answers of viewers and nonviewers to factual questions related to the content of the nature series *Australia Naturally*

Question	Viewers		Nonviewers		x^2	p
	Correct	Incorrect	Correct	Incorrect		
What does the word symbiosis mean?	17	96	11	106	1.225	0.191
What does a hermit crab do when it gets too big for its shell?	87	26	52	65	24.122	<0.001
Where does a snake's venom come from?	111	26	58	31	6.373	0.007
Why does the scrub turkey stick its head in its nest every day?	57	58	10	67	25.575	<0.001
Why do lizards and snakes lie out in the sun?	72	58	25	43	5.471	0.017
What does the word venomous mean?	87	50	44	45	3.822	0.050

But the programmes were not successful in teaching children what they should do when bitten by a snake, because most of the Australian children sampled already knew.

The fact remains that children can acquire factual information concerning both the meaning of words and the life-habits of animals from simply watching entertainment-orientated programmes on television. Since it will be later reported that children maintain that they use television sets in order to learn, it is useful at this juncture to note that this motive is realistically based as a consequence of viewing, even though the programmes they watched were shown on a commercial channel in entertainment viewing time (7.30 pm).

ACQUIRING MORE MORAL VALUES FROM ENTERTAINMENT PROGRAMMING ON TELEVISION

Having shown that children do learn from television, the next stage in the argument is to demonstrate that children also acquire those values which are, so to speak, "given off" in standard entertainment programming. Our examples here are again drawn from the *Australia Naturally* series.

In most contemporary nature programming the naturalist usually has an axe to grind in favour of conservation, and this is certainly true of *Australia Naturally*. In this instance we were able to check with the producer that the range of answers given to the children did accurately reflect the value position adopted in the programmes.

Table 2 Comparison between viewers and nonviewers in answers to the questions: "What important message was Bob Hardy, the presenter, trying to tell you in . . . (the various programmes)?"

Questions	Viewers		Nonviewers		x^2	p
	Agree	Disagree	Agree	Disagree		
After a programme on rainforests						
Rainforest trees keep the soil good and rich	75	43	13	64	39.085	<0.001
Should be careful where we cut down rainforest trees	38	80	8	69	11.120	<0.001
Should not cut down any rain-forest trees	71	47	12	65	36.085	<0.001
Should chop down all rain-forest trees and make farms	4	114	3	74	0.000	1.000
After a programme on estuaries and mangroves						
Should not spoil forests and rivers with pollution	57	36	8	78	49.994	<0.001
Should be careful not to upset the delicate balance of nature	66	27	14	72	51.874	<0.001
Estuaries and mangroves are beautiful	27	66	4	82	16.885	<0.001
It does not matter if we throw rubbish in the river	5	88	2	47	0.444	0.505

Results, which are summarised in Table 2, show that viewers had acquired the moral values "given off" by the programmes.

It is clear from the table that children acquired the moral messages in the *Australia Naturally* series with more force than they acquired factual information, since differences between viewers and nonviewers are larger in Table 2 than in Table 1. Moreover Table 2 shows that the more simple minded messages such as "chopping down all rainforest trees" and "it does not matter if we throw rubbish in the rivers" were already part of the children's conservation repertoire. It was only the more complex messages which the programmes conveyed – such as "being careful not to upset the delicate balance of nature" and "care should be taken when cutting down rainforest trees" – which were imbibed by viewing the programmes.

The argument advanced here is therefore that if children can acquire the moral messages of nature programmes simply by viewing, where these messages are reasonably clear cut and evident, then how many more "moral messages" are acquired from more diffuse everyday television programming about which research as yet has had little to say? This issue will be readdressed later in this chapter.

MECHANISMS BY WHICH LEARNING IS ACTIVATED WHEN VIEWING EVERYDAY TELEVISION PROGRAMMES

Communication research has recently involved cross-national comparisons of audience reactions. In a recent issue of the *International Journal for Political Education* (1981) researchers from many countries evaluated the impact of the programme *Holocaust* in their home countries. Audiences around the world reported that they had learnt about the Nazi's "final solution" from the series and that many audiences were inspired by the programmes to find out more about these events. Paletz (1981) wrote that in the Federal Republic of Germany "The program seems to have been seen not so much as imbedding factual incidents in a fictional plot but as a historical recreation, even actuality, represented in the lives of two families" (p. 3). In Australia we discovered with a sample of sixty undergraduate students (Noble and Osmond, 1981) that not only were the portrayed events seen as true fact (with no less than two-thirds reporting that the events seen in *Holocaust* were true to history) but also that the programmes had conveyed some understanding of why the Germans had engaged in the final solution to the Jewish question (only 11% reported no understanding at all). All of which, one notes, took place where there was no significant changes in ethnocentric attitudes, nor in attitudes to either Germans or Jews.

It is important not only to realise that mature adult audiences said they learned about Nazis and Jews from *Holocaust*, but to try to understand the mechanism by which such learning takes place. Noble (1975) proposed that viewers recognise regularly appearing characters on television and take answering roles to them, or in the *Holocaust* context extend to them sympathy and hope when they are encountered on the small screen in a viewer's living room. In other words, television works by reducing the scale of big events to one of personal relationships with the people which are seen on the television set.

Such an argument is not new. Horton and Wohl (1956) proposed a comparable process, whereby viewers parasocially interact with characters via what is termed "an illusion of intimacy" created between viewer and character on the set. My contention is that viewers who so engage with characters believe that what they view is reality and not fiction, as clearly happened in the instance of *Holocaust*. A further instance is demonstrated by the volume of medical mail which is addressed to the fictional Marcus Welby MD; no less than one thousand letters a week, according to Gerbner and Gross (1976).

Given the pattern of results mentioned above, it seems not only appropriate to give more attention in "learning" studies to characterisations on television but also to bear in mind the suggestion (Hyman, 1974) that mass media works on sentiments rather than on rational thought processes. Hyman directs our attention to the possibility that mass media work on audiences not by changing attitudes but rather,

... reverberate in our hearts and arouse such sentiments as pity ... I do not think we have the answer in past research because we have rarely formulated the question in quite the way that the concept, sentiment, would shape it. (p. 525)

Sentiments remain hard to define, which may be why they dropped out of the psychological literature in the late twenties. However my interpretation is that sentiments represent a fundamentally emotional response, typically involving two or more emotions, flowing together. In the *Holocaust* context, *The Economist* (1979) noted,

It took the screening on West German television of the American blockbuster about the persecution of Jews in Nazi Germany, "Holocaust", to do what scores of well-intended and often well-made documentaries, films, plays, and other broadcasts about Hitler's Germany never achieved: to provoke an urgent debate about a past which, even after 34 years, remains undigested and unredeemed. Soap opera it may have been, but to many Germans, "Holocaust", by scaling down ... six million murdered Jews to the more easily encompassed sufferings of a single family, made the final solution real for the first time. Shock, horror, disgusted surprise: these are salutory feelings about a gruesome period of recent history, and there is no point in being prim about the use of Hollywood methods if that is what it takes to achieve such an effect. (**270**, 7067, p. 14)

It seems Hollywood methods are in reality the means of appealing to sentiments rather than reason.

In our limited study of the learning from *Holocaust*, it was apparent that audiences reacted with their guts, rather than with their heads, the majority response being horror, disgust and amazement wrapped around with pleasure at watching the series, which was watched by more than half the Australian adult population. In addition we found evidence that the audience had reacted to the characters as people, and not just as Nazis. Dorf turned out to be the second most popular person in the series, second only to Rudi; yet at the same time Dorf was the second most hated character. These results forced us back to the literature of the early 1900s when Cooley (1929, p. 86) recognised that print was "the new communication spread like morning light over the world, awakening, enlightening, enlarging". At the same time he recognised the limitations of print,

We may read statistics of the miserable life of the Italians and Jews in New York and Chicago; of bad housing, sweatshops and tuberculosis; but we care little more about them than we do about the sufferers from the Black Death, unless their life is realised to us in some human way, either by personal contact, or by pictures and imaginative description. (pp. 88–89)

As *The Economist* editorial noted *Holocaust* achieved what scores of well-made documentaries about Hitler's Germany never achieved: sufferings of a single family made the "final solution" real for the first time.

Results from this study suggest that it is precisely the soap-opera nature of *Holocaust* which has provoked an urgent debate about Nazi Germany; pictures alone do not convey the message. Cooley's views would have been more accurate if he had combined pictures with personal contact, to give the confirmation encountered in soap-opera. Results suggest that it is through people relating to the very human characters in *Holocaust* that the sentiment of disgust, pity and horrified amazement is evoked.

Thus when considering social learning from everyday television programming, we are not talking about television as a surrogate for a school lesson; the unique medium of television must be regarded for what it is. We suggest that the mechanisms by which audiences say they "learn" about the world from television are quite unlike those by which facts are learnt in school. Our deduction is that most learning takes place from particular characters (a theme to be picked up when we consider how children use television both to develop concepts of morality and to acquire social skills not yet within their repertoire of experience), and from the feelings which the soap-opera genre engenders in its audiences.

Learning social skills from characters on television

Before shifting back to the above topic two observations must be reported to assist in the logic flow. The reader must first know that children say they watch television in order to learn, more than they say they watch television to pass or kill time. Amongst others to be reported later, the Greenberg (1974) study, which asked 180 nine- to fifteen-year-olds to first write an essay entitled "Why I like to watch television", found that two major content categories in answers were "to learn about things" and "to learn about themselves". When these, along with other more negative, reasons such as passing time or forgetting problems were incorporated in a questionnaire completed by 726 comparably aged children, learning emerged via factor analysis as the single most important and coherent factor, accounting for not less than twenty percent of the variance. To put this finding in perspective, the second factor, containing answers defined as "habit", which included reasons such as watching for mere enjoyment, interest, habit or for no specific reasons at all, accounted for only fourteen percent of the variance.

The second observation relates to the theoretical perspective outlined at the beginning of this review. If learning as a concept, particularly before the advent of mass education, implies finding a way along a track or path, then it seems self-evident that the way one learns to know the way along that path is through familiarity with that path, either direct or through the vicarious experience of another person. It is proposed here that television characters provide that surrogate experience which allows the viewer to know how to behave in social situations which he/she has seen on television but have not been encountered first hand in real life.

Again, when it is recalled that what we mainly watch on television is people, and when it is remembered that we mainly watch these people in a social context, it is clear that viewers may learn how to relate to people in real life from observation of how certain television characters relate to similar people on the screen. This would seem to be what Greenberg's children reported learning. Indeed Brown (1976) reports one child as saying "Before we had television I always wanted to go to live in a big city; now I'll stay where I am because I've seen what cities are like on the telly" (p. 134). This happily reverberates with a direct quotation from Herzog (1944) with reference to daily radio serials, "It helps you to listen to these stories. When Helen Trent has serious trouble she takes it calmly. So you think you'd better be like her and not get upset" (p. 53).

SOCIAL LEARNING FROM THE *HAPPY DAYS* SHOW

Back in 1976, when *Happy Days* was at the zenith of its popularity, Noble and Noble (1979) decided to try and discover how teenagers were "using" that programme. The idea of a "uses and gratifications" study came from overhearing teenagers discussing the programme amongst themselves. Initially therefore we set out to do two things – first to sit down and talk with young teenagers about the programme and the part it played in their lives, and second to try and discover what the producers intended to relay in the programme. This first objective was easily accomplished, and in group discussions teenagers reported that they did indeed learn how to behave towards other people from observing characters in the show. Not that they slavishly imitated the characters, rather they learnt by benefiting from the characters' experiences. On the basis of what they said, a questionnaire was constructed designed to measure *inter alia* the types of learning that *Happy Days* provided. The results are reported here.

Discovering what the producers were trying to do with the show was difficult until we found a publication by Herz (1976) entitled *The Truth about Fonzie*. The book "quotes" directly from Marshall, one of the producers, who describes Fonzie's philosophy, namely "He feels that if you want to do something in life you must do it. Even if the others make fun of you – if you enjoy doing something, don't pay attention to them. This is not necessarily 'do your own thing', but 'be your own person' " (p. 24). Such, it appears is what is meant by being "cool". More significantly when the scriptwriters revealed where they obtained the subject matter for their scripts, the ideas concerning learning about social relationships outlined above were very much in evidence. They admit, "it's entertainment . . . One reason entertainment exists is to show life at a heightened level. The relationships on *Happy Days* could appear in real life. The relationship between Richie Cunningham and his father is very important. On the show father is not always right" (p. 19).

Moreover the situations depicted in the show come direct from the real life experience of the scriptwriters. "When we have writers' meetings, we don't discuss ideas so much as incidents from our own lives. People think ideas come from imagination. Actually experience is one of the best places to get ideas. We sit in writers' meetings and talk about being young. We talk about things which happened to us" (p. 24).

Understandably the writers maintain:

The "boy–girl" relationship is always a big thing with young people. I don't think any boy was ever born knowing how to handle girls . . . These are things you have to learn . . . That's one thing we've tried to get across in the show through all the characters – that it's painful to grow up. We make a comedy out of it – but young viewers know someone else has been there before. (p. 25)

Given these intentions and results from preliminary group discussions, it was no surprise when we analysed results for a sample of 136 twelve- to seventeen-year-olds in Melbourne, Australia, to find that these adolescents overwhelmingly reported that they had learnt a great range of social behaviours from *Happy Days*. Results are reported in Table 3, as is the specific question asked along with the range of forced choice answers.

Results, which are reorganised in terms of most to least learning, clearly show that adolescents report that they had learnt how to be "cool" from *Happy Days*, and more significantly how to relate to the opposite sex and to friends. They also report that they learnt how to be "spunky"

Table 3 Amounts of social learning evident from answers to questions about *Happy Days*: "How often have the *Happy Days* characters shown you what to do/how to act in the following situations?"

	Very often		Often		Not so often		Never		Total
	N	%	N	%	N	%	N	%	
Be cool	76	57	15	11	14	10	29	22	134
Ask for a date	54	40	26	20	13	10	41	31	134
Where to go with friends	51	38	34	25	23	17	27	20	135
Be spunky	45	35	28	22	21	16	36	28	130
Things to do with friends	47	35	36	27	28	21	24	18	135
Popular with people	47	35	35	26	22	16	30	22	134
Not to be square	47	35	19	14	18	13	51	38	135
Behave on date	45	34	30	22	22	16	37	27	134
Be yourself	41	30	37	27	16	12	39	29	133
Get on with parents	38	28	48	26	19	14	30	22	135
Get own way	36	27	30	23	24	18	43	32	133
Wear on date	32	25	15	12	33	25	50	39	130
Become part of a gang	31	23	28	21	31	23	44	33	134
Do when you feel shy	23	21	32	24	33	24	42	31	135
Say no to people	27	20	27	20	36	27	44	33	134
Treat younger brothers and sisters	26	19	33	25	42	31	33	25	134
Behave at McDonalds	24	18	12	9	12	9	84	64	132
Jive/dance	23	17	32	24	35	26	42	32	132
Put people down	20	15	18	13	34	25	62	46	134
Do well at school work	14	10	11	8	31	23	78	58	134

(Australian slang for sexy). As Greenberg found they also reported that the show had shown them "how to be themselves". Predictably however, *Happy Days* had not shown sampled adolescents how to do well in school, nor how to jive/dance, nor how to put people down. These latter results are doubly important since they go some way towards establishing the validity of the findings.

When girls answers were compared with those of boys, it was not surprising that what the adolescents said they had learnt was appropriate to their sex roles. More boys than girls reported that they had learned how to ask for a date ($t = 4.308$, $p < 0.001$), and how to be popular with other people ($t = 3.891$, $p < 0.001$). However in areas such as relating to parents ($t = 1.094$) and learning how to be themselves ($t = 1.456$), there were no significant differences between the sexes, notwithstanding the male dominance of the casting in *Happy Days*.

Of greater theoretical interest are the results that take age into account. Among the twenty possible behaviours which could be learnt, there were no less than twelve significant correlations with age. In each case, as anticipated, results showed that it was younger adolescents who reported the greatest learning, presumably since older adolescents had already put into practise many of the behaviours on the list. Thus the correlation between age and saying one had learned how to ask for a date was highly significant ($r = 0.29$, $p < 0.001$). Similarly the correlation between age and learning "how to be spunky" was highly significant ($r = 0.25$, $p < 0.002$).

Among areas where social learning was not related to age were Fonzie's philosophy of "how to be cool" ($r = 0.12$, $p > 0.07$), "how to be yourself" ($r = 0.13$, $p > 0.07$) and two important relational behaviours – namely "how to get on with parents" ($r = 0.13$, $p > 0.06$) and "how to behave on a date" ($r = 0.10$, $p > 0.13$). Thus it would seem that teenagers do learn familiarity with various situations simply by watching other people cope with such situations on the television set. Indeed as the script writers imply, learning how to cope with fathers is something which all teenagers have to learn. Equally, learning to express and be yourself is a skill which, firstly, is not specifically taught elsewhere, and secondly, is not something that is learned solely in the early teens.

Yet the fact that so much of what was learned was age-related suggests that adolescents used *Happy Days* for anticipatory socialisations into roles which they envisaged they would have to enact in the future. It is precisely this surrogate experience with events likely or imagined to be encountered which constitutes social learning from everyday television. As Cooley (1929) said, media provide audiences with opportunities for "imaginative sociability".

DIMENSIONS OF SOCIAL LEARNING PROVIDED BY EVERYDAY TELEVISION

It is now time to complete the catalogue of "uses" which children find for television, as reported by Greenberg (1974).

One can report these uses as follows: (*a*) learning – both social and about oneself – contributing approximately 20% of the variance); (*b*) habit (14%); (*c*) relaxation or calming down (14%); (*d*) arousal or being stirred-up (13%); (*e*) to forget about real-life problems (13%); (*f*) to pass/kill time (5%), and (*g*) because of boredom (5%). No-one was more surprised at the saliency of the learning motive than was Greenberg; he had expected the more negative uses to outweigh the positive ones. When Rubin (1977) conducted a similar study with American children he found the same basic dimensions in children's uses of television, except that the pastime and habit motives were merged into one single factor, which was sufficiently large to challenge the learning factor for its position as the number one motive.

There are many such lists in the literature. McQuail, Blumler and Brown (1972) suggest certain basic functions which television can serve. These include, on the one hand, diversion or escape from everyday life, and, on the other, surveillance, whereby the viewer scans the set to obtain information about society and the world. These authors also suggest that many people watch television to establish personal relationships with the media characters, as substitutes for real-life companionship. At other times the viewer attends to the mass media to establish a personal identity, using the media to explore reality, to establish personal frames of reference and to reinforce viewers' existing values.

Given the fact that the research literature indicates that there is at least one positive reason for watching television for every negative one, Noble and Freiberg (1980) decided to explore what it was that children meant when they reported that they watched television in order to learn.

One hundred and six children aged from six to twelve years were asked to write down the names of television shows they most loved and hated, and to say why. Fifty children were subsequently interviewed both alone and in groups to flesh out the simple reasons given, such as because they were funny, exciting or scary, or simply in order to learn.

When we explored what the children meant by learning from television, it seemed that there were at least five different components in such learning. Each of these is very briefly illustrated below. First, rightly or wrongly, children reported in interviews that they learnt social mores (or customs, or acceptable behaviour in the wider society) from everyday television programmes. For instance, children said that *Starsky and Hutch* "shows the robbers cannot get away" (boy aged nine) or the *CHiPS* "stops people from driving fast" (boy aged eight). Such shows seem to reinforce the belief that law and order are for the public good and that law breakers will be brought to justice – at least as far as the child audience is concerned.

Secondly it appeared that some children, like adults, do watch television in order to keep an eye on the outside world – for surveillance. Such children, admittedly a minority, tended to watch news-type programmes for this purpose. At that time Skylab both threatened to and actually did fall out of the sky over Australia, and one eleven-year-old girl reported watching the news for Skylab, "in case it's over you and the DC 10 plane crashes. I feel like I'll never go in a plane". Another child said, "there should be more shows that show you how to act in an emergency" (girl, aged eleven).

The third component is relatively distinct to a country like Australia which is attempting to assimilate new migrants. When one bears in mind the fact that one in four Australian children is either a migrant or comes from a migrant family, it is not surprising to find that migrant and Australian children alike repeatedly report learning about the Australian past from television. Given the fact that Australian productions now hold their own with imported television programmes, and moreover are exportable, it was no surprise to find children saying "*The Sullivans* (a prime-time Second World War family serial) is so true. Dad said he was in the war and it was like that in New Guinea. The show is like it really was. Some of it is overdone . . . but it really did happen like that . . . You learn about Australian history and how they went away to other places" (girl, aged eleven). Equally "*Against the Wind* (a recreation of convict Australia) is gruesome, like whipping people. I feel like, as in *Roots*, they should be set free" (girl, aged ten). For the same programme "it shows how hard it was. It's different than school, they teach just about other countries, not early Australia, and *Against the Wind* gets more into peoples' lives . . . You find out about the past and they have the right sort of scenery" (girl, aged twelve). This latter comment is particularly illuminating since it highlights the differences between school-based learning and the learning which children report television provides. To refer back to the *Holocaust* findings, television conveys experiences which school-taught facts do not.

The fourth type of learning referred to anticipatory socialisation, which has already been described. However a couple of verbatim comments will illustrate what is meant here and allay the concern that forced choice-type answers of the sort used for the *Happy Days* study may be putting false words into childrens' mouths. One twelve-year-old primary school girl said of *Glenview High* (about an Australian high school), "It's good because it shows you toughies and what it is like at high school. The things they show really happen sometimes and will sort of help me when I go to high school". Equally the popular American programme *Family Affair* "tells you things that do happen in other families" (boy, aged nine).

Finally, much of what was learnt from everyday television programmes was seen as relevant to the self image of the viewing children – as reported in other studies. Thus of *Family*, an eleven-year-old girl said "it shows how Buddy sometimes fights with her parents. You can learn how to treat your parents from it". And in like vein a lengthy and illustrative quote from another eleven-year-old girl concerning *Family Affair*,

It shows what could happen to you and it can bring you close to other people, they (the children) sit and talk to Uncle Bill and he does anything for them. My parents would too, but it shows you what to do if you get into trouble . . . The things that happen in the show could happen to you and they show examples of what to say to mum . . . It's good for parents too, *they should watch and see how to treat you.*

Quantifying the incidence of learning

The results reported above were obtained in the first pilot year of a two-year study (Noble and Freiberg, 1980) funded by a commercial channel, ATN7, Sydney, which was designed to discover what children liked and disliked about television. The second year of the study involved sampling 300 children aged six to eleven from a high and a low social economic status area of Sydney. For the most part childrens' verbatim answers from the pilot studies were employed, along with open-ended answers in a structured questionnaire/interview approach, designed to ascertain how salient the learning motives were *inter alia*, and also to provide some information on the characteristics of the children who most frequently reported such "learning" from television. The frequencies are reported in Table 4.

Table 4 also reports on the frequencies of selected "uses" which children find for everyday television, rated on a different scale based on the one developed by Greenberg. As can be seen, surveillance motives were in a minority but nevertheless were more important than the nonmotive of habit. Overall, however, over half the sample reported that they learned each of the components outlined above, with the exception of surveillance proper.

A substantial majority agreed that television does teach things which are not learned in school. The most encouraging finding was that seven in ten agreed that television made them think about things. This result happily reiterates the findings obtained in the previously reported *Holocaust* study.

Learning in relation to background characteristics

Having ascertained sample frequencies for "learning" motives, discriminant analysis was applied to these data in order to discover which types of children were most inclined to report that they learned from everyday television.

There were no significant differences between boys and girls for any of the statements listed in Table 4, but social economic status differences were very much in evidence. Children from white-collar families were less inclined than their blue-collar counterparts to watch predictably because "it was a habit" ($F = 9.59$, $p < 0.002$), were much less likely to

Table 4 Frequency of various types of learning (in percentages) from everyday television for a sample of 300 six- to eleven-year-old Sydney children. Statements labelled B for question "How often do you watch television . . ."

Types of learning/statement used	Strongly agree/agree B: a lot	Neither agree nor disagree B: a little/ not much	Strongly disagree/disagree B: not at all
Stimulation: TV is good because it makes you think about things	70	14	16
Social mores: Some shows tell you things like "robbers cannot get away" and "you shouldn't drive fast"	64	15	21
Surveillance: TV tells me about important things you don't learn in school	64	15	21
B: To find out what's going on in the world	38	46	16
Australia: Australian TV shows are good because they teach me how people were treated long ago and what people went through	69	18	13
B: To learn about Australia	47	40	13
Anticipatory socialisation: I like watching TV because it shows you what could happen and how to behave	55	16	29
About the self: TV shows about families help me to behave at home and to talk to my parents	56	17	27
Control B: Because it's a habit	35	39	26

watch "to find out what it's like to be grown up" ($F = 34.53$, $p < 0.001$), and were less likely to watch television "to know how to behave at home and how to talk to my parents" ($F = 13.99$, $p < 0.001$). These results should not cause surprise, given the fact that social economic status in Australia is closely related to how long people have resided in the country, and confirm findings from the *Happy Days* study that newer migrants are more likely to look towards television to learn about the new culture than are established residents. White-collar children were much more likely to agree that "educational shows are good because they show you how to make things" than were blue-collar children ($F = 9.00$, $p < 0.005$).

Moreover, analyses undertaken to ascertain whether children watched either mainly the public broadcasting service (Australian Broadcasting Commission, ABC) or mainly the commercial channels, while contaminated by socio-economic status, also showed marked differences. Thus ABC viewers were much less likely to watch "to find out what it is like to be grown up" than were commercial viewers ($F = 29.07$, $p < 0.001$); equally ABC viewers disagreed more often with the statement that "television tells me about important things you don't learn in school" ($F = 10.51$, $p < 0.001$) and watch "to learn about Australia" ($F = 9.80$, $p < 0.002$). On the contrary commercial viewers were more likely to report that they watched because it was a habit ($F = 23.99$, $p < 0.001$). Again the picture which emerges is fairly predictable, more advantaged children seem to receive in their homes that stimulation about the world which disadvantaged children can probably only obtain from television. More importantly, those critics who maintain that televiewing is escapist and nonfunctional are likely to be well-heeled and unable to understand needs which the less articulate, but majority, section of the audience, both bring to and have satisfied by television.

Finally, analyses by age reveal that the learning provided by television is soon outgrown by the developing child. (Schramm *et al.* reported as long ago as 1961 that television viewing only provided a brief head start in vocabulary for six-year-old viewers, which was not sustained in older children.) Analysis was run to compare three age groups – namely six and seven, eight and nine, and, ten and eleven years. As far as watching "to find out what it is like to be grown up", agreement steadily declined with increasing age (6 and 7 mean = 2.17, 8 and 9 = 2.66, 10 and 11 = 2.97, $F = 8.55$, $p < 0.001$). Similarly agreement steadily declined with age for "I like watching television because it shows you what could happen to you and how to behave" ($F = 6.88$, $p < 0.001$), and for "watch television to learn how people in different places live" ($F = 5.15$, $p < 0.01$). Thus the components of learning labelled as surveillance and anticipatory socialisation were strongly related to age. As far as learning about the self was concerned there were no significant differences related to age, nor for that matter to any of the background characteristics herein reported. Indeed a U-shaped pattern of means for the motive "television shows about families help me to know how to behave and how to talk to my parents" was observed – with both youngest and oldest children agreeing, and the eight- and nine-year-olds disagreeing but not significantly so ($F = 0.29$, $p > 0.75$). This latter result corresponds with the U-shaped pattern observed by Hawkins (1977) in terms of the ages during which television is seen as accurately reflecting the reality of outside society.

In sum, it is worth stressing that while there were class differences in learning for anticipatory socialisation and about the self, channel differences in terms of learning about Australia, and age differences in terms of surveillance/anticipatory socialisation, there were no significant differences relating to background variables in terms of stimulation and social mores. Given the fact that so many comparisons were insignificant, we may tentatively conclude that "learning" as a motive for viewing is reasonably universal amongst the majority child audience.

Morality in relation to television viewing

By this stage in the analysis it is clear that children do learn from everyday television programmes, and that such learning does not rely on the process of imitation, but is rather mediated by the contiguity between people, places and events seen on television and those which the child either encounters, or thinks will be encountered in his/her life space. To reiterate this case, the results from a pilot study (Noble and Crabtree, 1981) which seeks to understand the relationship between the child's actual morality and his television exposure patterns show just how complex a phenomenon children's learning from television really is.

Wolfe and Fiske (1949) attempted to find out why children of that era read so many comic books. They discovered that young normal children used some comics as a means of successfully coping with reality by projecting, on to comic characters, those desires which parents found unacceptable. Thus *Bugs Bunny* was singled out by the children as "a tricky bunny who got away" with the things they wished they could get away with. Thirty years later, when talking to children in the pilot stage of the above "learning study", we found "Bugs Bunny is a bit tricky . . . I would like to do tricky things like Bugs and frighten my brother" (girl, aged seven). "Bugs is tricky. He plays tricks on people – like when the hunter comes, he pretends he's a lady. He gets away with it". Question: "Do you do tricky things like that?" Answer: "I don't, but I would like to, like Bugs. He gets away with things" (boy, aged eight). It seems children's media needs have not changed greatly!

In order to discover how children related to Bug's tricky morality and the morality of other television characters, 158 children aged from eight to eleven years from two schools in Armidale were asked to complete two questionnaires on two separate occasions, after completing the previously mentioned *Australia Naturally* questionnaires. The first questionnaire was designed to discover the level of the child's real-life morality, and the second to elicit television exposure, identification, and imitation of selected television characters as well as to ascertain the perceived morality of the same characters.

REAL-LIFE MORALITY

Dealing first with real-life morality, after two years of experiment in this area, nine morality stories were devised, including Piaget's question about stealing food for a hungry friend, which were deemed relevant to televiewing. Three questions were asked after each morality story. These results were factor-analysed by varimax solution to discover the structure of children's actual morality. Briefly, the results confirmed the hypotheses that, (*a*) one factor would index *selfish*, as opposed to, *altruistic* behaviour. The strongest correlations with this largish factor (48% of the variance) were for a story in which a boy and his friends collect bottles for the deposit money. These bottles are cleared out of the shed by one boy's father who collects the money and gives it to his son. The dilemmas concern whether the boy should keep the deposit money or share it with his friends. However it was clear from inspection that this factor also measured children's ideas about suitable punishments for various bad behaviours.

The second factor (34% of the variance) seemed to index hypothesis (b) *obedience to the letter rather than the spirit* of the law. Piaget's question relating to the boy who steals food from a shop for a poor hungry friend related most strongly to this factor, but analysis also showed that questions concerning obedience to rules, as well as to laws, were very much a part of this factor.

The third factor, which was small (18% of the variance), seemed to index hypothesis (c) *obedience to authority*. Questions concerning whether a boy should obey his parents' demand to go to a neighbour's house because they have to visit a sick relative in hospital; or keep his promise and go and help his friend finish building a canoe for a big race the next day loaded most strongly. However from inspection it appeared that suitable punishments for disobedience were also part of this factor. The validity of the factors was examined by looking at the correlations between factor scores and age, sex, hours of television watched and channels viewed. Scores representing *selfish–altruistic* morality were found to be only significantly related to age, with older children being more altruistic and younger ones more selfish. Scores representing *obedience to the letter rather than the spirit of the law* correlated significantly with all the background measures. Girls, public channel viewers, lighter viewers, children with white-collar backgrounds and, perplexingly, younger children obeyed the spirit rather than the letter of the law. Finally children deemed to be *obedient to authority* watched less television than those who were not.

TELEVISION MEASURES

When measure of exposure, behaving like and liking to be like television characters were examined, results indicated that Bugs and Beau Duke were most often watched and Basil and Fletcher least often watched by the sampled children. Children reported that they most often behaved like Ponch and Beau and least often like Sue Ellen, J.R., and Fletcher. In terms of which character children would most like to be like Bugs, Beau and Ponch headed the list with J.R. and Sue Ellen well and truly at the bottom of the pile.

PERCEIVED MORALITY OF SELECTED CHARACTERS

As far as the perceived morality of the characters was concerned, children rated J.R. as clearly the most *selfish*, followed some way off by Basil Faulty, Sue Ellen, Bugs Bunny and Hulk. The least selfish characters were, in order, Ponch, Laura, and Leela. Most *obedient to authority*, even if friends were hurt, the children reported J.R. (recall here the salience of Daddy in *Dallas*), Laura and Sue Ellen, while the characters rated least obedient to authority, in order to help friends were Fletcher, Basil, Porky Pig, Bugs Bunny, Hulk, and Beau Duke.

Finally in terms of who *obeyed the law*, even if it hurt friends, Ponch (*CHiPS*) appeared clearly, as expected, as the character who most embodied this sentiment. Few other characters were close to him, but Laura, Joanie, Buddy and Nicholas (mainly child actors) were all seen to obey the law more often than break it. On the contrary Beau headed the list as the character most often likely to break the law to help his friends, followed in turn by Hulk, J.R., Sue Ellen and Fletcher.

This condensed picture shows, first, that children answered these questions in a meaningful way, secondly, that there are a great variety of moralities on display by television characters, and thirdly, that there was sufficient variation in these data to warrant relating these scores to children's actual morality.

HYPOTHESES

Based on the Wolfe and Fiske study, the hypothesis was advanced that children who were still in an obedience to the *letter of the law* morality stage would want to be like *Bugs Bunny*. Since he embodied those desires the children were not allowed to express equally, we thought children who regularly watched Bugs would be likely to *obey authority* in real life, and would project their repressed desires onto Bugs. We also thought children's perceptions of Bugs' morality would be highly selective with children with differing real-life moralities seeing his morality in different ways.

RESULTS

Bearing in mind Wolfe and Fiske's and our finding that normal children outgrew Bugs in favour of super-hero characters by about ten years, the design of the present study was to partial out the effects of age, sex and social economic status to see if correlations between the child's actual morality and his television exposure measure remained after these variables had been eliminated. In terms of exposure, regular viewing of Bugs was correlated with real-life *obedience to authority* ($r = 0.10$, $p = 0.10$, $N = 158$). To put this result in perspective, regular viewing of Fletcher (*Porridge*) was correlated with real-life disobedience to authority ($r = 0.16$, $p = 0.02$). The correlation for Bugs was not much reduced by partialling out the above variables ($r = -0.107$, $p = 0.13$, $N = 120$), nor was the correlation for Fletcher ($r = -0.14$, $p = 0.06$). It appears therefore that the child's real-life morality does in part determine the television he chooses to watch regularly, and project upon his desires to be tricky. When children's real life *obedience to the letter, rather than the spirit of the law* were examined, results showed that children who wanted to be like Bugs were those who obeyed the letter of the law, ($r = 0.21$, $p < 0.005$). After partialling a near-significant correlation remained ($r = 0.14$, $p < 0.06$). Again, it seems that Bugs is chosen on the basis of wish-fulfilment as the character who does what the children would like to do. (In fact the results are remarkably similar for *Porky Pig*.)

Finally, correlations were examined between real life *selfish* versus *altruistic* morality. Here is transpired that children who said they behaved like Bugs were most definitely altruistic, $(r = 0.13, p < 0.06)$. After partialling this correlation was even more significant $(r = -0.17, p < 0.04)$. Thus we may conclude that there are very strong relationships between the actual real-life morality of the children and the way they relate to television characters, although this methodology does not permit generalisations about the direction of the influence.

SUMMARY

Children who regularly watch *Bugs Bunny*, it appears, are obedient to real-life authority and children who rigidly obey the law want to be like Bugs. One can see in both cases that Bugs is chosen most probably because the children project their desires onto him, since to express them in real-life would result in punishment. These we take to be the sentiments of the children's quotations reported above.

Moreover it appears that only children who are permitted, or allow themselves, to behave like Bugs exhibit the advanced morality of altruism rather than selfishness. One may predict that it is only when morality has matured that children no longer need to project unacceptable real-life desires onto Bugs.

SELECTIVE PERCEPTION

The next stage was to see if perceptions of Bugs' morality were selective. To this end, correlations were run between real life *obedience to the letter/spirit of the law* and measures of whether or not Bugs was perceived as *selfish*. Dealing first with those children that project unacceptable desires onto Bugs, that is, those who are obedient to the letter of the law, we found after partialling that such children perceived Bugs as doing what was best for other people $(r = -0.22, p < 0.01)$. Conversely, children with the more advanced morality underlying obeying the spirit of the law perceived Bugs as being selfish, that is, usually doing what was best for himself. This result we take to be an example of projection, whereby less morally mature children see Bugs' morality as mature, and, of course, *vice versa*. Once children are maturely prepared to obey the spirit of the law, they in turn see Bugs' morality as relatively immature.

Moreover, when the perceptions of Bugs' morality were examined for those children who were able to behave like Bugs, those who were altruistic in real life, the results showed, after partialling, that only children who displayed the advanced morality of altruism in real-life perceived Bugs as prepared to disobey what people in charge said in order to help friends $(r = 0.25, p < 0.005)$. Given the fact that these are correlations, the converse applies to children who were less morally advanced and were selfish in real-life. They in turn saw Bugs as prepared to do what people in charge demanded even if friends were hurt by so doing. The results establish that the perceived morality of Bugs, and for that matter other television characters, is most definitely selective. Such morality is in the eye of the beholder. In the latter instance, perceived morality appears to buttress the existing value position of the child, as was previously suggested by McQuail *et al.* (1972).

BIRD'S-EYE-VIEW OF THE TELEVIEWING STYLES OF ALTRUISTIC AND SELFISH CHILDREN

Because space does not permit a fuller exposition of these findings, a bird's-eye-view of the relationship between exposure to television characters and the age related selfish–altruistic form of morality in real-life will be briefly reported. No less than six out of a possible fifteen correlations remained significant after partialling out the effects of age, sex and socio-economic status. Analysis showed that *altruistic* children regularly watched J.R. (*Dallas*), ($r = -0.16$, $p < 0.05$), Leela (*Dr. Who*), ($r = -0.19$, $p < 0.02$) and Fletcher (*Porridge*), ($r = -0.17$, $p < 0.03$). Conversely, children with *selfish* morality in real life regularly viewed Ponch (*CHiPS*), ($r = 0.20$, $p < 0.02$), Beau (*Dukes of Hazard*), ($r = 0.29$, $p < 0.001$) and Nicholas (*Eight is Enough*), ($r = 0.16$, $p < 0.045$). The fact that these relationships remained after partialling out three background variables suggests that there are indeed very strong associations between the characters which children seek out on television and the maturity of morality displayed in real-life. Since we have already noted that children perceive J.R. as selfish on the screen, and yet he is attractive to nonselfish viewers, it looks as though J.R. is closely observed and acts as a model of "how not to behave" – or as Winch (1958) phrased it, for "negative identification". Television characters thus seem to be critically observed for contiguity between them and the child's life space, rather than imitated.

Conclusions

Much of this chapter has assumed that the reader watches television programmes which are popular with the child audience. Unfortunately it has been my experience that not all researchers are necessarily familiar with such shows, but it is by now clear that I feel obliged to be familiar with these programmes to conduct the types of study reported herein. The phenomenological approach, which has invariably been adopted as the first stage of all the reported studies, dictates that the researcher be so familiar. The morality results in particular may not mean a great deal to the reader unfamiliar with the selected television characters.

This caveat aside, the results reported here do amplify what children mean when they say they learn from television and begin to clarify some of the processes which underpin that learning. Learning is multifaceted, and two main processes seem to be at work. First, television relays experience mainly through feelings. As Jung pointed out long ago, the world can be apprehended just as effectively through feelings as through rational thought. Indeed, he argued that individuals eventually have to learn to use both modes of knowing, in order to understand the world. In broad terms, learning through feelings from television supplements the learning through thoughts which school provides. Secondly, television often provides the impetus for children to find out more about the events depicted in various programmes. However, my guess would be that it only provides temporary stimulation, which needs to be followed up in greater depth by scanning material from other media. Thus many children have told me that it is easier to read a book after seeing a televised version, since viewing helps in providing visual images which make the book come alive.

The more I think about this area, the more the theory of continuity (literally, "touching") seems appropriate as the mechanism by which children learn from television. That is children are not discriminant imitators of what they have seen on the screen – had they imitated J.R. from *Dallas* there would be cause for concern – but seem to learn by looking for common elements between what is seen on television and what they know to be relevant to their own lives. In other contexts contiguity could be dangerous. I have long argued that the "worst" form of violence on television is on newsreels. The recent riots in Britain seem like carbon copies of the Ulster variety, most probably because both the targets for the aggression and the weapons of aggression in English cities are contiguous with the targets and weapons of Ulster as seen in the living room over the past ten years. I fear it is my view that researchers, partly because they need to use scientific methods to "prove" their case, usually opt for simpler, rather than more complex processes. Thus it is that contiguity has been overlooked in the television learning arena. Hopefully, however, the learning studies reported in this chapter demonstrate that it is worthy of further study.

While writing in a contentious frame of mind I should point out that it seems doubly valuable that several of the learning studies here have relied in part on producer intentions. So often the researcher takes television as it is, without thought for its creation. Producers, I suspect, have long understood the way the medium does and can work in attracting audiences. It is my view that the producers of *Australia Naturally*, *Holocaust* and *Happy Days* all set out to entertain and inform. The results show that both objectives, not only entertainment, were reached.

To return to the beginning, learning, in my view, is finding a way along a path. Inhabitants of English inner cities have learned about the path of violence from what they have seen on television happening elsewhere in the United Kingdom. In the past, society's traditions have been passed down via folklore, but today television makes the past the present and relays some of society's mores instantaneously. All the same, many groups claim that they are not adequately represented on television. Such concern seems to amount, as Gerbner and Gross (1976) put it, to "symbolic annihilation", and seems to indicate that television relays only those mores that are understood by the majority of viewers. In this respect, television today is like the folklore of yesterday: the audience's ability to feel or touch such groups on television in itself demonstrates that some do feel that audiences can learn from television.

References

Brown, J. R. (1976). Children's uses of television. In J. R. Brown (ed.) *Children and Television*. Collier Macmillan, London

Cooley, C. H. (1929). *Social Organisation*. Scribners, New York

Gerbner, G. and Gross, L. (1976). Living with television: The violence profile. *Journal of Communication* 26, 173–199

Greenberg, B. S. (1974). Gratifications of television viewing and their correlates for British children. In J. G. Blumler and E. Katz (eds) *The Uses of Mass Communications*. Sage, California

Hawkins, R. P. (1977). The dimensional structure of children's perceptions of television reality. *Communication Research* 4, 299–320

Herz, P. (1976) *The Truth about Fonzie*. Scholastic Books, New York

Herzog, H. (1944). What do we really know about day-time serial listeners? In P. Lazarsfeld and F. Stanton (eds) *Radio Research 1942–43*. Duell, Sloan & Pearce, New York

Horton, D. and Wohl, R. R. (1956). Mass communication and para-social interaction. *Psychiatry* 19, 215–29

Hyman, H. H. (1974). Mass communication and socialisation. *Public Opinion Quarterly* 37, 524–40

Jung, C. G. (1921). *Psychological Types*. Princeton University Press, Princeton, N.J.

Katz, E. and Foulkes, D. (1962). On the use of the mass media as 'escape': Clarification of a concept. *Public Opinion Quarterly* 26, 377–78

Maccoby, E. E. (1951). Television: Its impact on school children. *Public Opinion Quarterly* 15, 421–4

Maccoby, E. E. (1954). Why do children watch television? *Public Opinion Quarterly* 18, 239–44

McQuail, D., Blumler, J. G. and Brown, J. R. (1972). The television audience: A revised perspective. In D. McQuail (ed.) *Sociology of Mass Communications*. Penguin, Harmondsworth

Noble, G. (1975). *Children in Front of the Small Screen*. Constable, London

Noble, G. and Crabtree, S. (1981). *Television and Childrens' Morality*. Mimeo. Armidale, Australia

Noble, G. and Creighton, V. M. (1981). *Australia Naturally – Childrens' Reactions*. Mimeo. Armidale, Australia

Noble, G. and Freiberg, K. (1980). *Australian Children's Uses of Television*. Mimeo. Armidale, Australia

Noble, G. and Noble, E. (1979). A study of teenagers' uses and gratifications of the *Happy Days* show. *Media Information Australia* 11, 17–23.

Noble, G. and Osmond, C. (1981). *Holocaust* in Australia. *International Journal of Political Education* 4, 139–50

Paletz, D. (1981). Introduction to reactions to *Holocaust*: Special issue. *International Journal of Political Education* 4, 1–4

Rubin, A. M. (1977). Television usage, attitudes and viewing behaviours of children and adolescents. *Journal of Broadcasting* 21, 355–69

Schramm, W., Lyle, J. and Parker, E. (1961). *Television in the Lives of Our Children*. Stanford University Press, Stanford

Winch, R. F. (1958). *Mate-Selection: A Study of Complementary Needs*. Harper, New York

Winn, M. (1977). *The Plug-In Drug: Televison Children and the Family*. Viking, New York

Wolfe, K. M. and Fiske, M. (1949). Why they read the comics. *In* P. Lazarsfeld and F. Stanton (eds) *Communication Research 1948–49*. Harper, New York

Researching the Effects of Television Advertising on Children: A Methodological Critique

Marvin E. Goldberg and Gerald J. Gorn

Introduction

For more than a decade, the issue of television advertising directed at children has kindled the thoughts and actions of the people and institutions that affect children's lives. Parents must deal daily with their children's television viewing habits and purchase-related behaviors. Schools have begun to incorporate into their curricula, materials developed to guide children in effective utilisation of television, including television advertising (e.g. see chapter 10 of this volume). Broadcasters and advertisers have developed self-regulatory codes to moderate the amount and nature of advertising to children. (In the US, the National Association of Broadcasters, *Television Code*, 1978; in Canada, the Canadian Association of Broadcasters, *Broadcast Code for Advertising to Children*, 1980). Politically, the issue has been hotly contested, with hearings at the US Federal Trade Commission, where several hundred interested parties presented briefs, resulting in a 1981 decision not to develop a general rule limiting television advertising directed at children (Federal Trade Commission, 1981). By contrast, in Quebec, Canada, the legislature recently enacted a Consumer Protection law, two clauses of which make it illegal to advertise to children (Quebec National Assembly, 1978).

Researchers also have been involved, assessing the role television plays in children's purchase behavior. A recent bibliography (Meringoff, 1980) references over 500 entries, of which over 200 are research studies. In the midst of the discussion regarding this issue, the role of the researcher would ideally be to provide a basic understanding of the processes associated with children's exposure to television advertising and their subsequent purchase-related behaviors. Given a reliable and valid data base, debate within the society might then proceed on a more factual, less emotional basis.

Researcher's methods are often assumed (by the layman in particular) to be sound, unbiased and without prejudgment. Yet any research paradigm has inherent methodological weaknesses and built in biases (Kuhn, 1962). It is the purpose of this chapter to consider many of the typical problems faced by researchers in addressing the issue of children and television advertising. Attention will be focused, for the most part, on the strengths and weaknesses of the two major research approaches that have been utilised: survey research with its reliance on self reports and correlation, and experimental research with its reliance on laboratory settings and internal controls.

In the following section, problems in measurement are examined. Some of the problems are, in theory, common to both surveys and experiments, yet tend to be encountered more with one approach than the other. For example, problems such as the inability or unwillingness of respondents to answer questions accurately emanate from the limitation of self-report measures. Both experiments and surveys may fall victim to these problems, yet they may be encountered less frequently in experiments where other types of dependent measures (such as behavior) are also used. Another example is the problem regarding the appropriate level of specificity associated with dependent measures. Surveys in the area of children and television advertising have tended to use overly general, or less specific dependent measures, while experiments have tended to use overly narrow or more specific measures. The trade-offs associated with these types of problems are discussed.

In the third section, problems in the assessment of causality are considered. Causality is defined in its broadest sense, incorporating issues of external as well as internal validity. The very way television exposure is defined in either surveys or experiments makes it either more or less likely that a causal link will be found between television exposure and various purchase-related behaviors. For the survey researcher, the respondent "self-selects" (i.e. identifies himself) as having been exposed or unexposed to a given television commercial; or more generally, as a "light" or "heavy" television viewer. Inasmuch as a series of factors are likely associated with a predilection for, say, heavy or light television viewing, the survey approach makes it inherently difficult to tease out these associated "third factor variables", complicating efforts at establishing causality (in particular, internal validity).

The experimenter typically manipulates exposure/nonexposure, thereby facilitating efforts at establishing causality, at least within the context of the experiment. Yet a variety of problems confront the experimenter in establishing external validity, or "causality in the real-world". Basic among these is the likelihood that some of the subjects who are "forced" to see a television commercial within an experimental condition and hence defined as "exposed" would never choose to do so in the "real-world". Possible strategies for minimising, if not overcoming, the difficulties both surveys and experiments have in establishing causality are considered in this section.

In the fourth section we deal with perhaps the central issue with regard to child directed television advertisments: can children effectively "arm" themselves against television commercials i.e. can they process the messages accurately and develop cognitive defenses to defend against the messages? There has been a tendency for most survey researchers to answer in the affirmative, and for most experimenters to answer negatively. It is suggested that methodological biases may contribute to these different research outcomes and related perspectives. In particular, survey researchers tend to document a *generalised* understanding and distrust of commercials *in response to* survey questions. Experimenters tend to document the inability of children to cope *in situ* with powerful television commercials directed at them. Much in question is children's ability to *activate* the defense mechanisms they may have acquired *vis-à-vis* television commercials.

The final section. The biases that emanate from a survey research or an experimental perspective are not limited to a single issue but rather interact pervasively with the researcher's theory building and testing. In the area of television advertising and children, some researchers have utilised a consumer socialisation model which may be contrasted with the "effects" model employed by experimenters. The values implicit in both the methodologies and theoretical paradigms adopted are considered in this section.

Problems in measurement

Whether used in surveys or experiments, the self-report is likely the most common method of assessing the influence of television advertising on children. The validity of the replies, is often unquestioned, perhaps because the respondents (mothers, or the children themselves) have no apparent motive to distort their responses. Yet a number of potential biases in self-reports regarding television viewing and children's purchase related behavior ought to be examined.

THE INABILITY OF SUBJECTS TO ANSWER QUESTIONS ACCURATELY

Very young children, in particular, may be unable to recognise and hence accurately report on questions regarding the relative importance of various sources of purchase influence. For example, in one study, preschool children recalled considerably less about a commercial (brand name, product attributes, theme) than did older, third grade children (Ward, Wackman and Wartella, 1977). This poorer memory for television messages is likely to extend to purchase-related incidents in general, making reliance on at least young children's self-reports tenuous.

In two separate studies, younger children cited stores as the most frequent source of ideas for toys (Ward *et al.*, 1977; Atkin and Reinhold, 1972). Yet at least in the latter study, the children's mothers did not agree, citing television as the major source of product information. Given their poorer memory of the details of particular commercials, young children are unlikely to be able to sense and accurately report on the interaction between the product information they retain from television commercials and their recognition of these products in the store. It would seem inappropriate to rely on the responses of five- or six-year-olds, given their limited understanding as to how selective perception operates so that in a store with hundreds or thousands of toys, they come to focus on one or two they know about and are interested in. In their minds, the salience of the store situation is likely to predominate. Much more subtle and likely to escape their awareness is the role television commercials play in alerting them to and teaching them about these toys. If by contrast, older children come to have some understanding and report more accurately regarding this subtle interaction between television and store, this does not necessarily imply that television becomes more important as a source of information about products as children grow older. These comparisons can yield useful developmental insights only insofar as the younger and older children who respond are both equally *capable* of providing valid responses and are motivated to do so.

THE UNWILLINGNESS OF SUBJECT TO ANSWER QUESTIONS HONESTLY

Even if they can understand the question and remember events as accurately as adults, children, like adults are sometimes motivated to distort their answers. This problem may be more pronounced for older children, once again making any cross-age comparisons moot. Thus as Rossiter (1979) notes:

Children's increasingly negative attitudes toward TV advertising do not mean much . . . they merely acquire an adult like attitude against TV advertising as a social institution; an attitude which bears little relationship to advertising's actual effects. (p. 232)

Thus while older children may more readily deny being influenced by television, than do younger children, these differences can be due to a learned social desirability bias in the responses of the older but not the younger children. As but one example, the Reilly Group Inc. (1973) asked children: "When you see a TV commercial for a food, would you like the product more if it had a premium or was nutritional?"; only one-third of those eleven and over said "a premium", compared with more than half of those under ten. This is but one situation in which the social desirability of the (nutritional) response may well have been more salient for the older children and contributed as much to the between age differences as any true attitudinal differences *vis-à-vis* premiums and nutrition.

The problem of a social desirability bias in self-reports is an important one as well, when mothers are the survey respondents: Rossiter and Robertson (1975) indicate that compared to the responses of their children, parents (1) underestimated their children's total viewing time; (2) overestimated the amount of coviewing; (3) overestimated parent–child interaction and (4) underestimated their children's susceptibility to commercials. The researchers conclude: "a social desirability bias underlies the general pattern of idealized reporting of television control by parents" (p. 308).

PROBLEMS IN THE WORDING OF QUESTIONS

In almost any questionnaire the way the question is posed can make for considerable differences in the nature of the responses. This is true of questions posed to children regarding the influence of television commercials upon them. Thus Ward et al. (1977) indicate that in response to the question "what is a commercial?" only 10% of kindergarten children mentioned the persuasive aspect of advertising. In response to the question "what do commercials try to do?" 22% reported that commercials try to get them to buy products. When asked "what does this commercial for (product X) want you to do?" considerably more (about half) responded that the commercial wanted them to buy or try the product (as reported in Wartella, 1980). In this series of questions it is not immediately evident whether the first question is too obtuse and leads to an underestimate, or, by contrast, the third question is too leading and hence represents an overestimate of children's awareness of the commercial's selling intent. It is clear, however, that the way the question is asked can make an enormous difference in the nature of the responses.

Occasionally a question is framed in a patently biased fashion. Thus in two studies the following question was posed to children: "How much does it bother you when they stop the program to show the commercial". The skewed nature of the question might immediately lead one to interpret with caution the overwhelming number of responses in the "a lot" and "sometimes" categories (97% and 89% in the two studies). Indeed only 23% of the mothers of these same children indicated they felt that disruption of the program was a concern for their children (Atkin, 1975).

THE LACK OF SITUATION/TIME SPECIFICITY ASSOCIATED WITH QUESTIONS

One of the problems with the latter type of question is that, even if it is framed in a more balanced fashion it tends to be very general in nature, and as such the opinions elicited are unlikely to be predictive of behavior (Fishbein and Ajzen, 1975). Making these questions considerably more situation specific, perhaps by using behavioral preference measures, ought to be considered. Thus for example, to measure children's level of annoyance at commercials interrupting a program, one might construct a situation in which they were led to believe they would actually be watching a videotape of a favorite program of theirs and that the experimenter had two videotaped versions: the original with commercials for GI Joe and Strawberry Shortcake dolls, McDonalds and Burger King, etc. And a second version with these commercials deleted. The children would then be asked which version they would prefer to watch. The immediacy and specificity of the choice suggests a greater degree of validity to the responses. Indeed, once reminded of the specific commercials they would be missing one might hypothesise that a considerable percentage of the children would opt for the commercial version of the videotaped program.

This type of situation-specific measure tends to be used more often in more narrowly focused experimental studies. For example, in one experiment with preschoolers, Goldberg and Gorn (1978) sought to measure children's predilection for material objects (toys) over persons, as a function of exposure to commercials for these objects. Children who had either seen or not seen a commercial for a toy were shown two boys, one holding the advertised toy, the other without it and were asked:

I can bring one boy to come play with you. I can bring this boy who is not so nice and you can play with him and his (toy) or I can bring this boy who is nice (pictured without toy). Would you like to play with the nice boy or would you like to play with the boy who is not so nice and his (toy).

While 70% of those in the unexposed control group opted for the nice boy, well over half of those who had seen the commercial opted for the not-so-nice boy and his toy. A second parallel question asked the children whether they would rather play by themselves with the toy or with their friends. Once again a significantly higher percentage (70% vs 30%) of those exposed to the commercial opted for the toy as opposed to their friends. Further, these results were maintained upon retest 24 hours later, suggesting that the responses were, in fact, reliable.

Of course these two behavioral preference questions sample only two situations from the universe of possible situations from the "materialism" domain, and as such reflect the opposite type of weakness relative to the overly general self-report questions discussed above. As is typical for measurement of any trait, a combination of both broad ranging and situation-specific measures is likely to yield the most reliable and valid indicator.

Just as experiments tend to include more situation specific measures than do survey/self-reports they also focus on a more immediate time frame. In a classic article, Hovland (1959) suggests that this differential focus on the immediacy or remoteness of the effects assessed can influence the magnitude of the effects obtained.

In the typical experiment, the time at which the effect is observed is usually rather soon after exposure to the communication. In the survey study, on the other hand the time perspective is such that much more remote effects are usually evaluated. When effects decline with the passage of time the net outcome will, of course, be that of accentuating the effect obtained in experimental studies as compared with those obtained in survey researches. (p. 9)

The immediacy of the (purchase) situation can be critical. For example the harried mother denying her toddlers requests in the supermarket situation may provide an estimate of the level of parent–child conflict quite at odds with her response in the quiet of her home after her children are asleep in the evening.

Another example of how the timing of a study can affect the salience of an issue (and hence the accuracy of measuring it) was a Christmas-time assessment of the link between exposure to toy and game commercials and children's related requests (Robertson and Rossiter, 1976). In tracking children's requests over the month prior to Christmas, a five percent increase in the proportion of requests for advertised toys and games was noted; this was consistent across different age levels. In addition, heavy viewers were significantly more likely than light viewers to request advertised toys and games. Clearly the timing of this study was one of its major strengths.

IMPRECISION IN CATEGORIES OF MEASUREMENT

Often, the low correlations found in self-reports may be the result of attenuation due to measurement error. Thus the weak relationships noted in one study (Ward et al., 1977) between level of exposure to television commercials and purchase requests may have been due to the way in which both variables were measured. Frequency of children's purchase requests was measured by giving children a list of products and asking how often they asked their parents to buy each item: Children could choose one of four responses: "often – once a week or more"; "sometimes – once a month or more"; "not too often – less than once a month"; or, "never".

Even assuming children (especially the kindergarteners questioned) had: (1) a reasonable degree of self-awareness of their behavior in this regard; (2) a reasonably accurate memory of their behavior and (3) an understanding of the time frames associated with each category, the very broad nature of these categories was likely to have attenuated any true relationship between frequency of purchase requests and any dependent variable.

The independent variable, described by "frequency of exposure to television commercials", was obtained in the following fashion:

We asked mothers how many hours their children spend watching television on an average weekday after school, on an average weekday evening and on Saturday. We summed these measures with appropriate weighting as an index of the children's exposure to commercials. (Ward et al., 1977, p. 140)

Even assuming mothers tried to be as accurate as possible and the social desirability of their responses played no role, the concept of an *average* number of hours extrapolated to obtain a total is bound to contribute to considerable measurement error. In addition, child-related products were probably not advertised during some of these blocks of time, further contributing to measurement error. Furthermore, using total hours of exposure to television does not account for the considerable variance among children with regard to the number of hours of noncommercial public television they view. That this makes a considerable difference was highlighted by one study (Galst and White, 1976) which distinguished between the total number of commercial or noncommercial hours of television the children viewed per week. The total number of hours of commercial television watched per week was positively and significantly related to the children's number of purchase influence attempts; however the purchase influence attempts was not related to the *total* number of hours of television watched per week. This lack of correspondence was due to the influence of the hours of noncommercial television watched per week.

Correlations have been low (0.20 and less) even when mothers have been asked to collect daily diary data on their children's viewing habits (Henderson, Kopp, Isler and Ward, 1980). One likely contributor to measurement error under these circumstances is that parents spend very little time viewing television with their children; as low as 20% coviewing is reported on Saturday mornings (Nielsen, 1975) and so they often must approximate the amount of time they believe their children were in front of the set.

Further, children may or may not attend to the television even if they are in the television room, and in the absence of coviewing, mothers cannot report on this. Research suggests relatively high levels of inattention. Ward *et al.* (1977) reported that when mothers were trained to systematically observe their five-to-twelve-year-old children watching television, and recorded their level of attention as full, partial or none, full attention was noted only about two-thirds of the time; Bechtel, Achelpol and Atkin (1972) report even lower levels of attention, with one-to-ten-year-olds watching commercials only 40% of the time they were on. Even more important perhaps is the fact that attention levels are not equivalent across age groups. In in-home or naturalistic environments preschool and elementary school children tend to pay more attention to commercials than do older children (Wartella, 1980). Thus, in comparing younger and older children, any drop in the correlation between total number of viewing hours and purchase requests, may be due to the fact that number of viewing hours is a less appropriate measure for older children than it is for younger children.

In contrast to these relatively low correlations of 0.2 and less, Galst and White (1976) report a correlation of 0.52 between level of exposure and purchase requests. This is likely due to the precision with which they measured the child's level of exposure to television commercials. They asked children to come see three half-hour programs including commercials at three sittings, each at least a week apart. To be able to view the television material the children had to press a button. The amount of time the button was depressed both for the commercial and the program was recorded for each child. These same children were followed by a researcher throughout a trip to the supermarket with their mothers and the number of purchase influence attempts was recorded. A correlation of 0.52 was obtained between the time each child had depressed the button to view the commercials, and the number of purchase influence attempts he made in the store.

While this observational study benefits from its precision of measurement, it is clearly a more arduous procedure, resulting in a minimal sample size (41 children). By contrast, as long as care is taken to consider its possible shortcomings, the survey self-report, represents a relatively economic and efficient way of collecting an abundance of data on healthy sample sizes. The need for these complementary approaches is self evident.

Problems in assessing causality

The very definition of exposure to a television stimulus tends to differ in surveys and experiments, and these definitions lead to different problems in assessing causality. For the experimenter, artificially manipulating exposure can lead to problems in external validity. For the survey researcher, allowing respondents to "self-select" into exposed or unexposed groups typically creates problems with the internal validity of the responses.

EXPERIMENTS: DEFINING "EXPOSURE" THROUGH LABORATORY MANIPULATION CAN LEAD TO AN EXAGGERATED ESTIMATE OF EFFECTS

As Hovland (1959) has noted, experiments tend to distort reality by creating experimental conditions in which individuals who typically would not expose themselves to the stimulus, suddenly are confronted by it:

. . . in naturalistic situations, with which surveys are typically concerned, the outstanding phenomenon is the limitation of the audience to those who *expose themselves* to the communication. Some of the individuals in a captive audience experiment would, of course, expose themselves in the course of natural events to a communication of the type studied; but many others would not. (p. 9)

Children range considerably in terms of the number of hours of television they view. While the average may be more than 3½ hours of television viewing per day or almost 26 hours per week (Nielsen, 1975) one study reports a daily range of 2–6 hours (Robinson and Backman, 1972) and a second (Lyle and Hoffman, 1972) found that 25% of sixth graders reported no viewing on a given weekday, while another 25%

viewed more than 5½ hours. Given this range of viewing, experiments exposing *all* children to a particular commercial, including those who might not typically view the commercial, might potentially exaggerate the relationship between exposure and a given dependent measure. As such, while the experimenter might establish a clear causal link within the experimental setting, the relationship might be said to represent the *potential* as opposed to the actual level of effects manifested in reality.

As one example, consider the link between the child's level of exposure to television commercials and his level of unhappiness when he is denied what is advertised. Low level television viewers who might not normally see the product advertised, nor ask for it and be denied it, may well display considerable discontent once exposed in the context of experimental (forced) exposure. However, this would not necessarily reflect the reality of their home situation. In this regard, Goldberg and Gorn (1978) asked preschoolers who had either viewed or not viewed commercials for a "Ruckus Raiser Barn" whether a boy (pictured watching television) was sad that he didn't get the Ruckus Raiser Barn or still happy because he could watch television instead? Almost twice as many (about 60% *vs* 35%) of those viewing the commercial *versus* those who did not, projected that the child was sad. Although perhaps not directly comparable, evidence from at least one survey suggests a somewhat lesser level of actual disappointment in the home. Only 35% of children who had had their Christmas requests denied, expressed disappointment (Robertson, 1979). Children were able to rationalise their failure to receive their requested gifts in a number of ways.

SURVEYS: SELF-SELECTION IN DEFINING EXPOSURE RESULTS IN ASSOCIATED THIRD VARIABLE PROBLEMS

If experiments have difficulty establishing external validity, surveys have equal difficulty establishing internal validity. Consider, for example, survey researchers' efforts to assess the link between children's level of exposure to television commercials and their level of purchase requests. Atkin's (1980) approach is typical. Children are asked to: (1) *define themselves* as heavy or light television viewers (i.e. "self-select" into one group or the other), and to (2) indicate their level of purchase requests for a wide variety of food products, toys and fast-food restaurants. Atkin reports that heavy viewers made about twice as many purchase requests as did light viewers. The problem is, however, that self-selection of respondents into high or low exposure groups allows for any number of third variable factors, minimally to moderate and maximally destroy, any relationship that is manifested.

For example, it might be argued that the level of parental yielding can contribute directly or indirectly to the child's level of television exposure. In a direct sense, children who anticipate success in their requests for products, may well want to scan their environment broadly to know what to ask for, and so will watch more television. Less directly, acquiescent or more generally permissive parents, may establish a home environment or climate, that encourages (or at least permits) heavier television viewing. Thus the child's expectations regarding the success of his request, which develops as a function of the degree of parental yielding, could determine both the level of television exposure as well as the number of purchase requests, making the television exposure–purchase request relationship spurious. Under these circumstances, variations in levels of purchase requests would more properly be attributed to the child's expectation of success as a function of his parents level of yielding, and not to high or low levels of exposure to television commercials.

RESOLVING THIRD FACTOR PROBLEMS AT THE EXPENSE OF
EXTERNAL VALIDITY: *EXPERIMENTS*

The line of reasoning presented in the previous section would suggest that a child's expectation of success may have as much to do with his level of purchase requests as does exposure to television commercials. To formally test this hypothesis, the authors designed two experiments (one with middle-class and one with lower-class boys) in which expectations of success and exposure were independently manipulated (Goldberg and Gorn 1974; Gorn and Goldberg, 1977). These studies demonstrate at once the strength of the experimental approach in assessing causality, along with its inherent weakness: a narrowness relative to the real-world phenomena it attempts to model.

The experimental design structured as independent: (1) the child's expectancy of success in obtaining an advertised toy; and (2) his exposure to commercials for a (new) advertised toy. While all of the children were given a description of the toy and very briefly viewed it in its package, only some got to see the commercial for it, within a cartoon program. To create levels of expectancy, the boys (in groups of 15) were told that there were either 1, 8, or 14 toys for the 15 of them and that the first 1 (or 8 or 14, depending upon the condition they were in) to solve a puzzle they would be given following the program, would win the advertised toy. Since the puzzle was essentially insoluble the situation was (as intended) a frustrating one for the child. The key variable was a measure of motivated behavior: the length of time the boys were willing to persist at this frustrating task. In both studies, those who were exposed to the commercial for the toy, worked significantly longer in efforts to obtain it, relative to boys who did not see the commercial for the toy. This main effect operated independently of the expectancy manipulation. With upper-middle-class boys a significant expectancy effect was also obtained. Boys who felt they had a better chance of obtaining the toy persisted longer than boys who felt they had a poorer chance. This expectancy manipulation did not have any effect with the lower class boys, apparently because any chance the boys had to obtain the toys within the experimental context was better than their real-world expectancies. No interactions were noted in either experiment. It would appear that the net result of these experiments was to establish a causal link between television exposure to an advertised product and a child's increased efforts to obtain it. Further, approaching the television situation with a high expectancy of obtaining the advertised product may also result in greater efforts to obtain it. The lack of interactions suggests, for example, that even children with less of a chance of getting a toy are still affected by the commercials; they do not totally ignore it, and conversely, those exposed to the commercial are not carried away by it but continue to recognise the constraints of their situation (their expectancies).

These experiments appear to have established causal links between the two independent variables and the main dependent measure in a way surveys cannot. Yet what of their external validity; their ability to estimate the extent of the causal link outside the laboratory? It is in this regard that the typical laboratory narrowness or artificiality of the above measures might be considered:

1. Exposure to the television commercial takes place in the children's recreation center. As familiar as they are with this setting, it is not their own home. They are watching, not by themselves or with one or two friends, but in a group of 15 boys.

2. The expectancy levels that were created may have been internally valid, but the results themselves suggest a problem with their external validity. The superpositioning of artificial expectancies on the real world expectancies of the low income boys, in particular, appeared to have failed. For them, having *any* chance to win the toy within the context of the experiment seemed better than their everyday chances of obtaining it. The experimental manipulation could not stand up to the robustness of the parallel real world factor.

3. The dependent measure, a child's persistence at a puzzle, intended as a proxy for his persistence in the face of his mother's (or father's) failure to yield, points to both a strength and a weakness of the experimental approach. The experimental paradigm did permit precise measurement of persistence in terms of a clear continuum of time; but if the internal validity of this measure is clear, its external validity is arguable. For example, one of many elements of potential artificiality in this measure relates to the nature of the feedback the child receives. Failure to solve the puzzle was a form of negative feedback, and likely led to a level of frustration, which in turn provided a measure of motivated behavior. By comparison, however, failure to convince a parent to buy a toy typically generates more than impersonal negative feedback. Typically the parent's denial of a purchase request will occur within the larger interpersonal context, and will often be accompanied by some specific type of rationale for the denial.

One ploy parents typically take in denying their child's requests is to try and distract the child by offering a preferred (cheaper, more creative etc.) alternative. An experiment by the authors suggests that this approach will be considerably less successful, with children who have been exposed to a commercial for the toy in question. Goldberg and Gorn (1978) showed a Ruckus Raiser Barn toy commercial to one set of preschoolers and not to a control set. It was suggested that the experimenter had spoken to the child's mother and that the mother had preferred "this tennis ball" (i.e. a clearly inferior object in the child's eyes) over the Ruckus Raiser Barn. The child was then asked which one he/she preferred. About 80% of the control group who had not seen the commercial (but who had seen the toy and were familiar with it) deferred to their mother's judgment. But only half of the group exposed to the commercial deferred to their mother's judgment.

Going further, Prasad, Rao and Sheikh (1978) experimentally manipulated the type of explanation a mother provided her child as she denied their request for an advertised product. They showed two commercials (one somewhat more attractive than the other) to eight-to-ten-year old boys. At the same time, they trained the boy's mothers to coax their children away from the advertised products, in either a power-assertive or a reasoning mode. The results suggest that the television commercials carried the day against the mothers in all but one condition. (Only when the less attractive television commercial was followed by the mother in the "reasoning" mode did the boys select an alternative unadvertised reward to any significant extent.)

The latter two experiments appear to serve as a partial bridge between the laboratory measure of children's persistence at a puzzle in order to win a toy, and in-home persistence in the face of parental denials of purchase requests. The cumulative efforts of these experiments appear to suggest that exposure to television commercials for a sought after product will make children more likely to override negative feedback or parental denials.

EXTENDING THE EXTERNAL VALIDITY OF EXPERIMENTS THROUGH BLOCKING

It might be argued that the external validity of the Prasad et al. (1978) study discussed above is limited, to the extent the mothers were trained to act counter to their typical mode of dealing with their child. An interesting way of countering this concern and extending the study of the relative effects of parental permissiveness and television exposure would be to "block" on the parental permissiveness factor. Children with parents who are assessed as either permissive or nonpermissive (using a self-report measure) could each be subdivided into groups either exposed or unexposed to television commercials and a factorial design established. While some loss of internal validity may result, the real benefit would be the direct assessment of television exposure alongside of a key external real-world factor.

An interesting application of blocking was used by Donahue, Meyer and Henke (1978) who showed McDonald's commercials to both black children and white children. The results suggest that the two groups perceived the commercials very differently. Relative to the white children, the black children saw the advertised food as better than the food they received at home, and they viewed the family portrayed in the commercial as relatively happier than their own family.

Another example of this approach of blocking on an external real-world variable is found in the literature on television violence's effects of children. Cline, Craft and Courrier (1973) examined the effects of exposing children to television violence by first assessing how much television the children watched at home. Subjects were then divided into low and high television exposure groups. Results suggested that high television exposure children were significantly less aroused when exposed to television violence in the laboratory than were low television exposure children. (The authors interpreted this as suggesting a general desensitisation to violence as a result of children's high exposure to television violence in the real-world.)

In the same way, children with histories of high or low exposure to commercial television could be subdivided into groups that are exposed or unexposed to a test commercial. This blend of utilising experimentally manipulated as well as external, television related, factors as independent variables can represent one way of dealing with Hovland's (1959) admonition that the experimental communications paradigm typically places very secondary emphasis on factors other than the impact of the single manipulated (television) variable. While field experiments are often suggested as a way of achieving greater external validity, they are often costly, time-consuming and cumbersome; as a result they are relatively rare. By contrast, blocking can be a way of maintaining the efficiency of the laboratory while at the same time introducing a degree of external validity into experimental research.

EXTENDING SURVEYS VIA TIME SERIES ANALYSIS AND MULTIVARIATE MODELLING

Given the problem of numerous potential third factor variables, it is unlikely that simple correlational efforts will be adequate to establish a degree of internal validity. Time series analysis is one way to help extend the survey research paradigm and address this problem. With a time series analysis, one would look for changes in children's responses as a function of changes in television advertising at different points in time. For example, in Quebec, one could assess changes in the level and pattern of children's food and toy choices as reflected in: cereal sales (sugared *vs* nonsugared); sales volume for various categories of candy and toys etc. both before and at various points after the elimination of television advertising to children. An equivalent change occurred when American television channels were first introduced into Newfoundland in the late 1970s via microwave (from Maine). Prior to this, only Canadian stations with very little advertising directed at children had been available. This presented an opportunity to consider the same type of changes in the level and patterns of children's food and toy choices, beginning with the introduction of cable and at various points as the diffusion of cable progressed (Barnes, Russel and Kelloway, 1978).

Multivariate modelling procedures may also help in disentangling third factor variables such as level of parental yielding from the relationship between level of television exposure and level of purchase requests. In the development of *a priori* models, survey researchers could carefully specify relationships between relevant independent and dependent variables. Equations could be written for each of these relationships. They could then be estimated simultaneously. The goodness of fit of the entire model or part of it could then be tested (Bagozzi, 1980). In this way survey research can both incorporate a wealth of moderating variables and test specific hypotheses.

This was the approach employed by Henderson *et al.* (1980). A linear equation was set up to predict purchase requests as a function of a number of variables (yielding, age, television exposure, family interaction, etc.). The impact of television exposure was negligible. A linear equation was also set up to predict parental yielding. As would be expected, the predictor variables were not necessarily the same as those used to predict requests. For example, the mother's consumer skills were incorporated into the yielding but not the request equation. Number of requests was used as an independent (predictor) variable in the equation to predict yielding, with yielding used in the equation for requests. (A two-stage least square procedure was adopted to deal with the mutual causation between requests and yielding in the two equations.)

Both the suggested extension of the survey research paradigm through the use of modelling procedures and the extension of the experimental paradigm through the use of blocking on external real-world factors represent two of several possible avenues for improving the complementarities of surveys and experiments. (For a related discussion see Comstock, Chaffee, Katzman, McCombs and Roberts, 1978, pp. 491–500). In this regard Hovland's (1959) conclusions are no less relevant almost 25 years later:

Integration of the two methodologies will require on the part of the experimentalist an awareness of the narrowness of the laboratory in interpreting the larger and more comprehensive effects of communication. It will require on the part of the survey researcher a greater awareness of the limitations of the correlational method as a basis for establishing causal relationships. (p. 14)

Neither is a royal road to wisdom but each represents an important emphasis. The challenge of future work is one of fruitfully combining their virtues so that we develop a social psychology of communication with the conceptual breadth provided by correlational study of process and with the rigorous but more detailed methodology of the experiment. (p. 17; for a parallel discussion see Cronbach, 1957)

Methodological biases: researching children's cognitive defenses *vis-à-vis* television commercials

A central issue with regard to television advertising directed at children, is whether or not child viewers develop effective cognitive defenses that "arm" them against television commercials. Survey researchers have focused on the generalised understanding children acquire regarding the nature and purpose of commercials. By and large they suggest that this understanding represents an adequate defense. Experimenters have tended to be more skeptical, suggesting that children might not *activate* these defense mechanisms when actually confronted by powerful television commercials. It would appear that the methodology employed by survey researchers and experimenters leads them to ask different questions in different settings, and so, to reach different conclusions.

SURVEY RESEARCHERS: CHILDREN ACQUIRE GENERAL COGNITIVE DEFENSES

Survey researchers frequently cite a developing distrust of television commercials as a child grows older as evidence that children can effectively process commercials and defend themselves against these messages. Gaines and Esserman (1981) indicated that almost all four-to-five-year-olds and all six-to-eight-year-olds can correctly identify a commercial as separate from a program. Further, close to two-thirds of the four-to-five-year-olds and 85% of the six-to-eight-year-olds were said to have "evidenced understanding of the purpose of a commercial". The rather generous definition of "understanding" used in this study highlights the question as to what the proper definition ought to be. Children who responded "commercials show you things you can buy" were coded as understanding the purpose of commercials (one quarter of all the children and one third of the youngest ones provided this response). While it is likely that young children do indeed regard television commercials as providing trusted friendly advice (i.e. "shows you things") this may well be evidence of a demonstrated lack of comprehension of the selling intent of commercials, rather than the reverse.

While one might question Gaines and Esserman's (1981) findings that almost two-thirds of four-to-five-year-olds understand the purpose of commercials, it seems clear that by age ten or eleven, most if not all children do understand the purpose of commercials, in the sense that they (1) can discriminate between commercials and programs; (2) recognise a sponsor as the source of the commercial message; (3) comprehend the intentionality or selling purpose of commercials and (4) understand the commercials are symbolic, scripted, filmed and unreal (Rossiter and Robertson, 1974; Ross, Campbell, Huston-Stein and Wright, 1981). Whereas just over half the first graders in one study recognised the persuasive intent of commercials, virtually all did, by the fifth grade (Robertson and Rossiter, 1974). Correlations between age and understanding of the purpose of television commercials range between 0.45

(Robertson and Rossiter, 1974) and 0.66 (Wackman, Wartella and Ward, 1977). These correlations are interpreted as evidence that even with cumulative exposure to television commercials (20 000 per year; Adler, Ward, Lesser, Merringoff, Robertson and Rossiter, 1980), children are better equipped to deal with them as they mature. Yet, it would seem too early to conclude, as some have, that children are not unduly vulnerable targets (Rossiter, 1979):

It is reasonable to regard these cognitive measurement results as evidence against the proposition that children become more "mentally" susceptible to TV advertising techniques as they are exposed to more TV advertising. (p. 50)

What one can conclude, is that by age eleven or so, *when asked* questions such as "(do) television commercials tell the truth" or "(can) you always believe what the people in the commercials say or do" (Rossiter, 1977), children have indeed learned the appropriate (adult) responses. The question remains, however, whether in face of the immediate and powerful stimulus that a commercial represents, children actually can or do *call on* their generalised distrust of commercials in processing and evaluating a particular message.

EXPERIMENTERS: DO CHILDREN ACTIVATE DEFENSE MECHANISMS THEY MAY HAVE ACQUIRED IN THE TELEVISION VIEWING SITUATION?

Perhaps because they come face to face with children as they watch the television screen, experimenters are typically unconvinced that children can effectively defend themselves against television commercials.

Ross *et al.* (1981a) suggest at least three ways in which children's information processing abilities are limited in dealing with television commercials:

Children are limited in: [1] their ability to apply conceptual knowledge to override perceptual impressions; [2] their ability to search systematically for relevant information, rather than responding to immediate stimulus salience; and [3] their ability to consider multiple properties of stimuli [e.g. visual and verbal] simultaneously. (pp. 330–1)

In a carefully executed experiment, they assessed the accuracy of judgment of children from kindergarten to sixth grade regarding the presence of actual fruit in three types of cereals and beverages advertised on television: "real fruit"; "nonfruit" and "artificially fruit-flavored" products. Showing the children these commercials resulted in greater accuracy (regarding the presence of actual fruit) in the "real fruit" and "nonfruit" conditions but greater *inaccuracy* for the artificially fruit-flavored product condition. These effects were noted regardless of age. In other words, commercials for artificially fruit-flavored cereals and beverages led older as well as younger children to be even more inaccurate than they had been prior to exposure, regarding the presence of real fruit in these products. The authors conclude that the strategies employed in these commercials are such that children's growing general skepticism notwithstanding, they are confused by the messages. *In the television viewing situation*, verbally, the words "real fruit flavor" appear to leave an impression of "real fruit" that stands in the children's memory as fact. Visually, fruit shown on the package or elsewhere, leads even older children to the conclusion that the product contains fruit.

Similarly, Ross, Campbell, Wright, Huston, Rice and Turk (1981b) reasoned that if an understanding of the selling intent and strategies of commercials and a corollary distrust of them is sufficient "defense", older children should be less influenced by extraneous information in a commercial, such as the use of a racing car celebrity or real racing footage in selling a toy racing car. In fact, manipulating the presence or absence of these extraneous factors in commercials shown to older and younger boys, resulted in a main exposure effect: the celebrity endorser and real racing footage were effective regardless of the age of the boys. No interactions were noted. Older children were no more accurate than younger children in processing the extraneous information presented in the commercials. The researchers concluded that at least in this television viewing situation, the older children failed to employ the television-related cognitive defenses they may have acquired. What children may "know" about commercials (or at least so respond in direct questioning) is not necessarily the critical dimension in assessing their abilities to cope with the persuasive messages coming at them.

Further evidence of the limitations of children's defenses *vis-à-vis* television commercials is provided by a study timed to cover both the beginning and end of the preChristmas television advertising barrage (Rossiter and Robertson, 1974). In assessing children's defenses the researchers distinguished between "cognitive and attitudinal defenses". Cognitive defense was based most centrally on the child's ability to understand the selling intent of the commercial. Children who disbelieved, disliked and were disinclined to try advertised products were said to have attitudinal defenses. Correlations were computed between children's cognitive and attitudinal "defenses" and their preference levels for television-advertised toys and games, before the peak of the preChristmas advertising period and at its conclusion, one week prior to Christmas. Initially, prior to heavy advertising, children with the strongest defenses to commercials selected fewer television promoted toys and games than children with weaker defenses. However after a month of heavy television advertising, the children's defenses were "ineffective predictors of preference". The authors term this a "defense override" and suggest that it is due to an increase in preferences for television-advertised goods among the children who initially had the strongest defenses. Thus, at least the barrage of Christmas advertising appears to be able to overcome any general defenses children may have acquired.

Roberts, Bachen and Christenson (1978) suggest that basic to the issue of whether children can adequately defend themselves against television commercials is whether they *counter-argue* as the message is ongoing. Does the child use an understanding of the purpose of a television commercial to actively counter-argue as he watches it? In effect, does he learn (as do adults) to *discount* the influence attempt because it is a paid commercial message? Roberts and Maccoby (1973) suggest four conditions are necessary for counter-arguing to occur: the viewer must : (1) recognise the persuasive intent of the message; (2) realise that persuasive intent implies biased information; (3) be involved enough to be *motivated* to engage in counter-arguing; (4) to have the ability, knowledge and experience to draw on as counter-arguments.

While there is as yet no direct evidence regarding the extent to which children counter-argue and at what age, these four conditions are likely to provide useful guidelines in facilitating relevant research. Thus for example: (1) when and how is the persuasive intent and biased information in a television commercial likely to be salient for a child viewer; (2) what (other) conditions may be specified that will lead child viewers to be motivated to counter-argue; (3) what constitutes sufficient knowledge and experience (with a product) so that the child has the ability and "ammunition" to counter-argue?

In one experiment, two 15-minute films demonstrating the selling methods and strategies used in television commercials were sufficient to lead even second graders to process commercials more skeptically (Roberts, Christenson, Gibson, Mooser and Goldberg, 1980). Statements such as "Bill Cosby knows more about Jello than most people" were presented to the children after all had viewed a commercial with Bill Cosby endorsing Jello, but only some had seen the treatment films. Those who had viewed either of the two films were considerably more skeptical both with regard to the specific commercials they viewed and with regard to commercials in general. It appears that films such as these represent one way that children can be provided with both the information and motivation to process television commercials more carefully and skeptically.

The interaction between methodology, theoretical paradigm and pervasive biases in research

The biases that may be a function of a survey research or an experimental perspective are not necessarily limited to a single issue such as the one considered in the previous section. In this section we shall consider how both the methodologies and theoretical paradigms researchers adopt are not necessarily value-free but may lead to pervasive biases both in the perspective taken and conclusions drawn. The contention is that: (1) both the consumer socialisation model adopted by many researchers, and the survey research methodology utilised, often leads to a conservative perspective, and (2) the "effects" model of the experimenter and his "interventionist" methodology often leads to an activist perspective.

CONSUMER SOCIALISATION, THE STRUCTURAL-FUNCTIONAL PARADIGM AND CONSERVATIVE BIASES

Much of the research that has considered television advertising's effects on children has fallen within the area termed "consumer socialisation", defined as: ". . . processes by which young people acquire skills, knowledge, and attitudes relevant to their functioning as consumers in the marketplace" (Ward, 1974, p. 2). More broadly, socialisation typically refers to:

An adult initiated process by which developing children, through insight, training and imitation acquire the habits and values *congruent with adaptation to their culture*. (Baumrind, 1980, p. 640; italics added)

Implicit in a socialisation perspective is that there is an ongoing functioning system into which people will integrate. Also implicit is that it is functional and adaptive to become integrated, i.e. socialised. In the present context, this implies that it is children who should learn to adapt to the system, rather than the system that should be adapted to conform to the limitations and/or needs of the child. With regard to television advertising and children, attention tends to focus more on how *children* should change, rather than how television should change:

The roots of . . . dissatisfaction (with children's TV commercials) are undoubtedly often a result of children's failure to critically evaluate advertising messages and their own product desires. To the extent that children can be taught to filter advertising claims more rigorously, evaluations of products should result in fewer "mistakes" in product purchases. (Ward *et al.*, 1977, p. 173)

The consumer socialisation perspective minimally is congruent with, and likely emanates from the sociologists structural-functionalist paradigm (as typified by Talcott Parsons, *The Structure of Social Action*, 1949). With its focus on the societal "system" and the functional nature of that system the structual functionalist paradigm has been described as a *conservative* theory. It:

[treats] the institutions of its own surrounding society . . . as given and unchangeable in essentials; proposes remedies for them so they may work better rather than devising alternatives to them; foresees no future that can be essentially better than the present, the conditions that already exist; and explicitly or implicitly counsels acceptance of or resignation to what exists rather than struggling against it . . . Functionalists are not pollyanas who see no fault in the status quo. But neither do they see any possibilities of a future significantly different from the present. (Gouldner, 1970, pp. 332–3)

The focus in the classical structural functionalist systems model is that of equilibrium and not change. The relatively minor adjustments that are needed can be instituted through the efforts of individuals, and need not invoke entire institutions. From a consumer socialisation and television advertising perspective, the *parent* can limit the child's amount of television viewing; the *parent* can be less permissive, the *child* can be taught to be more skeptical, the *advertiser* can adopt a more socially responsible attitude, etc. The implicit or explicit view is that the system is essentially viable and functional and that any changes that are needed are relatively minor and within the scope of individual action:

. . . our data suggest opportunities for marketeers to undertake positive programs which could effectively respond to many of the concerns of all but the most extreme individuals. (Ward *et al.*, 1977, p. 186)

Instead of a ban on advertising addressed to children, I have in mind using programmatic materials to teach children about advertising and products; instead of requiring advertisers to sponsor health or nutrition disclosure messages . . . I propose teaching children how to evaluate foods and establish good diets. (Ward, 1978, p. 135)

EXPERIMENTAL INTERVENTION AN AN ACTIVIST PERSPECTIVE

The consumer socialisation model typically assumes children will eventually learn to rationally evaluate the product alternatives available to them: "the child understands what he or she learns in most commercials, selects those products which are interesting and attractive and asks for them" (Rossiter, 1981, p. 232). Even if this rational model were correct, children could make optimal choices only where they were presented with the full range of alternatives available. From an economist's perspective, choosing from a skewed array would not, in theory, produce optimality.

Content analysis of television advertising directed at children suggests a highly skewed array of products: about 60% of all commercials during children's hours are for food related products, and about 70% of these are for highly sugared foods such as sugared cereals, cakes, cookies and candy (Barcus, 1980). For whatever reasons, foods such as fruits, vegetables or other breakfast foods are virtually unrepresented during children's television hours. Similarly, where toy commercials dominate, as during the fourth quarter (prior to Christmas), the heavy stress on the benefits of owning and playing with toys is, of course, nowhere counterbalanced by messages considering the benefits of alternative activities such as book reading or outdoor play.

It remains for cross-cultural studies to try and determine how children learn to make choices in countries where commercialism is not so pervasive and whether the range of options they have leads to different judgments regarding optimality in allocating their time and dollars. Within the context of North American society, however, we might ask how the skewed nature of the choices children face (for an average of three hours a day on television) affects their judgments. One nutritionist (Gussow, 1972) has suggested that just as heavy exposure to commercials for soda pop or beer leads adults to the refrigerator to open a Coke or a

Budweiser rather than to the sink to open the tap for water; so too, children's exposure to sugared snacks and cereals leads them to select these foods rather than the relatively unadvertised fruit, vegetables and other breakfast foods.

One experiment (Gorn and Goldberg, 1980) suggests that the very perception and definition of the choices available depends upon how broad and heterogenous an array is presented. Eight- to ten-year-old boys were first shown a number of commercials for two candy bars and were asked to indicate how different or similar (on a four point scale) the candy bars were. A week later they viewed a number of commercials for fruits and were asked once again to indicate how different or similar the two candy bars were. As predicted, the children evaluated the bars as significantly more similar after exposure to the fruit commercials than before. Making fruit more salient appeared to have broadened the children's frame of reference resulting in greater perceived similarity between the two candy bars. Thus the very definition of what constitutes a meaningful choice can be altered by changing the array of choices available.

It is the proposition that alternative "realities" can be examined by the active intervention of the experimentalist that leads him to a broader questioning of the ongoing system. In this instance the central research question that is posed is not simply whether children can adequately handle the choices presented to them, but rather should these choices be other than what they are presently. It is the experimentalist as interventionist that typically has raised the question: should television as one institution within the society be fundamentally changed? In its divergence from the consumer sociologist's perspective this approach parallels the views of the more activist sociologists of the late 1960s for whom "conflict" and not "system" represented a more accurate societal model (e.g. Marcuse, 1964). Friederichs (1970) goes as far as to suggest that these divergent paradigms were a reflection of the researcher's own self-image: the structural functionalist dominant in the 1950s, an era of prosperity, tranquility, conformity; the activist-conflict sociologists coming into dominance in the late 1960s and early 1970s, an era of conflict and "mass disaffection with the system". Change is far more central to the conflict model, and as Gouldner (1970) notes, change is seen as resulting from broad collective or institutional solutions (and not the narrower individual actions envisaged by the structural functionalist).

One broad ranging institutional change that has been discussed (Federal Trade Commission, 1981) is the wholesale elimination of television messages for highly sugared foods targeted at children. "What if", there were no television commercials for highly sugared foods aimed at children? "What if", there were an equivalent number of commercials for various fruits and/or Public Service Announcements encouraging children to moderate their intake of sweets? It is to these alternative institutional realities that a number of experimenters have addressed themselves.

In the first of two studies, the authors showed first graders either highly sugared or more wholesome snack breakfast food messages during a half-hour cartoon program (Goldberg and Gorn, 1978). They were then asked to indicate which foods they would want if their parents were away on a holday and the experimenter was babysitting for them. In this test situation, the children's snack and breakfast choices clearly reflected the type of food messages they had viewed earlier.

In a second study (Gorn and Goldberg, 1982) the same pattern of results held true for children's actual snack food choices in a summer camp setting. Each afternoon, for 14 days, during their "quiet hour" five- to eight-year-old children viewed a half-hour cartoon program into which was inserted $4\frac{1}{2}$ minutes of television messages for either high- or low-sugared snacks. Each afternoon for their snack they were then given actual choices of either high- or low-sugared snacks such as candy bars or fruits. With real choice behavior as the criterion, children's snack food selections reflected the type of television messages they had viewed.

Behaviorists involved in studying eating behavior have long recognised the importance of environmental factors. In efforts to combat obesity (e.g. Stuart and Davis, 1972) strategies center around making desired foods salient through their ready visibility and ensuring that undesirable foods are out of sight. The strategy of the behaviorists is to *change the environment not the person.* Exposure on television to one type of food or the other represents an important element of the child's environment. Daily exposure to a candy bar or a fruit may well be the equivalent of having the candy bars or fruit sitting on the kitchen counter. Given this perspective, it may be argued that it would be more effective to change the nature of the child's television exposure experience rather than trying to change the child.

SURVEY RESEARCH AND EXPERIMENTS: DIFFERENT METHODOLOGIES, DIFFERENT PARADIGMS

In the long run, in the "real-world" would a wholesale change in the nature of television food messages that a child viewed actually change his snack habits? The perspectives of the survey researcher and experimentalist likely lead to different conclusions, as Hovland (1959) notes:

The survey researcher . . . stresses the large number of simultaneous and interacting influences affecting attitudes and opinions . . . In contrast, the experimentalist frequently tends to view the communication process as one in which some single manipulative variable is the primary determinant of the subsequent attitude change. (pp. 14–15)

In the survey researcher's predilection for examining a large network of variables together with their interactions, is an implicit suggestion that reality is best explained by an array of influences and not a single causal factor. In a sense, this perspective is related to the structural functionalist's multifaceted system and his focus on the equilibrium of that system. Thus one might cite a variety of influences aside from television commercials (school cafeterias, vending machines, corner stores, parental eating habits and purchases, friends' eating habits, etc.) that would all have to be changed before one could realistically expect a child's eating habits to change. Ward *et al.* (1977) list 27 independent "family context variables" of which frequency of exposure to television commercials is but one. (The time and effort required to assess each of these factors is one indication that they are not lightly considered.) In Henderson *et al.*'s (1980) regression model where children's purchase requests were predicted, ten variables were included (of which "television viewing", was one). No single beta weight was above 0.30.

With the experimentalist, the implicit assumption is quite the reverse. While other influences (institutions, individual, etc.) may be controlled, measured as independent factors and found to interact with exposure to television commercials, the assumption is that television is "the primary determinant". (One could hardly justify the elaborate efforts that often enter into an experimental assessment of television effects if it were not for the assumption that television is likely to be the key factor.) In a sense, the perspective is related to the conflict sociologist's, with the simplifying assumptions that enable him to focus his energies on a single institution.

These differing perspectives are not unique to the television advertising and children issue. Experimentalists and survey researchers who examined the television violence and children issue arrived at equally divergent findings (Comstock *et al.*, 1978). Early in the 1940s and 1950s survey researchers tended to downplay the effects of television violence suggesting that what was most readily detected was a self-selection of viewers that merely led to a reinforcement of existing tendencies. By contrast, the experimental findings of Berkowitz and Bandura suggested powerful broad-ranging effects of television violence on children's aggressive tendencies.

Comstock *et al.* (1978) suggest that socio-political events played a considerable role in resolving this issue. With increased societal violence in the US throughout the 1960s (assassinations, war) and extensive viewing of increased levels of television violence, the educated citizenry developed a growing reluctance to accept the no-effects perspective. Together with more and more convincing research data (much of it generated by a special $1 million research program on television violence in 1969) the weight of scientific opinion moved, in the 1970s, to accept the likelihood that viewing television violence contributed in some way to aggressive behavior in society.

If socio-political events colored judgments regarding research related to the television violence issue (and, in fact, led whole generations of sociologists to adopt different paradigms), one might posit the influence of recent socio-political events on the children's television advertising issue. One possibility, is that given the termination of the US Federal Trade Commission's consideration of rule-making with regard to children's television advertising (FTC, 1981) at least some of the research interest in this issue may dissipate. Another possibility is that at least some research will focus on an assessment of specific alleged deceptions of particular commercials, following the guidelines now used by the FTC in considering the children's television advertising issue (FTC, 1981).

From the perspective of a cumulative science, however, it is perhaps self-evident that these more specific pieces of research will need to be integrated into a broader analytic framework. To encourage a balance of methodologies, we have tended to highlight the benefits of the experimental approach. In fact, it is to be hoped that the issue will continue to attract the research efforts of as broad a spectrum of social scientists as possible. It is only by incorporating a wide diversity of perspectives, as this book attempts to do, that researchers can succeed in building a robust general paradigm of children's purchase behavior and television's role in shaping that behavior.

References

Adler, R. P., Ward, S., Lesser, G. S., Merringoff, L. K., Robertson, T. S. and Rossiter, J. R. (1980). *The Effects of Television Advertising on Children: Review and Recommendations.* Lexington Books, Lexington, Mass.

Atkin, C. (1975). "Effects of Television Advertising on Children–Survey of Preadolescents' Responses to Television Commercials". Technical report, Michigan State Univ. *Reported in* Atkin, C. Effects of Television Advertising on Children. *In* E. L. Palmer and A. Dorr (eds) *Children and the Effects of Television.* Academic Press, New York

Atkin, C. K. (1980). Effects of Television Advertising on Children. *In* E. L. Palmer and A. Dorr (eds) *Children and The Faces of Television*, pp. 287–305. Academic Press, New York

Atkin, C. and Reinhold, C. (1972). "The Impact of Television Advertising on Children". Paper presented at the meeting of the Association for Education in Journalism, Carbondale, Illinois, August. *Reported in* Atkin, C. Effects of Television Advertising on Children. *In* E. L. Palmer and A. Dorr (eds) *Children and the Faces of Television.* Academic Press, New York

Bagozzi, Richard P. (1980). *Causal Models in Marketing.* John Wiley, New York

Barcus, F. E. (1980). The Nature of Television Advertising to Children. *In* E. L. Palmer and A. Dorr (eds) *Children and The Faces of Television.* Academic Press, New York

Barnes, J. G., Russel, B. A. and Kelloway, K. R. (1978). *The Socio-cultural Impact of Cable Television on the Community of St. John's Newfoundland.* Part IV: Analysis of first wave data. School of Business Administration and Commerce, Memorial University of Newfoundland

Baumrind, Diana (1980). New Directions in Socialization Research. *American Psychologist* **35**, 639–652

Bechtel, R. B., Achelpol, C. and Atkin, R. (1972). Correlates between observed behavior and questionnaire responses on television viewing. *In* E. A. Rubenstein, G. A. Comstock and J. P. Murray (eds) *Television and Social Behavior*, Vol. 4. *Television in day-to-day life: Patterns of Use.* US Government Printing Office, Washington D.C.

Canadian Association of Broadcasters (1980). *Broadcast Code for Advertising to Children*, 8th edn. Ottawa, Ontario

Cline, V. R., Craft, R. G. and Courrier, S. (1973). Desensitization of Children to Television Violence. *Journal of Personality and Social Psychology* **27**(3), 360–365

Comstock, G., Chaffee, S., Katzman, N., McCombs, M. and Roberts, D. (1978). *Television and Human Behavior.* Columbia University Press, New York

Cronbach, L. J. (1957). The Two Disciplines of Scientific Psychology. *American Psychologist* **12**, 671–684.

Donahue, T., Meyer, T. and Henke, L. (1978). Black and White Children's perceptions of television commercials. *Journal of Marketing* **42**, 34–40

Federal Trade Commission (1981). "FTC Final Staff Report and Recommendation. In the Matter of Children's Advertising 43" (p. 14). Washington D.C.

Fishbein, M. and Azjen, I. (1975). *Belief, Attitude, Attention and Behavior: An Introduction to Theory and Research*, Ch. 8. Addison-Wesley, Reading, Mass.

Friederichs, R. W. (1970). *A Sociology of Sociology*. Face Press, New York

Gaines, L. and Esserman, J. (1981). A Quantitative Study of Young Children's Comprehension of Television Programs and Commercials. *In* J. F. Esserman (ed.) *Television Advertising and Children Issues, Research and Fundings*, pp. 96–105. Child Research Service, New York

Galst, J. and White, M. (1976). The Unhealthy Persuader. The Reinforcing Value of Television and Children's Purchase-Influence Attempts at the Supermarket. *Child Development* **47**, 1080–1096

Goldberg, M. E. and Gorn, G. J. (1974). Children's Reaction to Television Advertising: An Experimental Approach. *Journal of Consumer Research* **1**, 69–75

Goldberg, M. E. and Gorn, G. J. (1978). Some Unintended Consequences of TV Advertising to Children. *Journal of Consumer Research* **5**, 22–30

Goldberg, M. E., Gorn, G. J. and Gibson, W. (1978). TV Messages for Snack and Breakfast foods: Do they Influence Children's Preferences? *Journal of Consumer Research* **5**, 73–81

Gorn, G. J. and Goldberg, M. E. (1977). The Impact of Television Advertising on Children from Low Income Families. *Journal of Consumer Research* **4**, 86–88

Gorn, G. J. and Goldberg, M. E. (1980). TV's Influence on Children: The Long and The Short of It (1979). *In* G. J. Gorn and M. E. Goldberg (eds) *Proceedings of The Division 23 Program*, 88th Annual Convention of the American Psychological Association, Montreal, Canada

Gorn, G. J. and Goldberg, M. E. (1982). Behavioral Evidence of The Effects of Televised Food Messages on Children. *Journal of Consumer Research* **9**, 200–205

Gouldner, Alvin, W. (1970). *The Coming Crisis of Western Sociology*. Basic Books, New York

Gussow, J. (1972). Counternutritional Messages of TV Ads Aimed at Children. *Journal of Nutrition Education* **4**, 48–52

Henderson, C. M., Kopp, R. J., Isler, L. and Ward, J. (1980). "Influence on Children's Product Requests and Mother's Answers: A Multivariate Analysis of Diary Data". Report No. 80–106. Marketing Science Institute, Cambridge, Mass.

Hovland, C. I. (1959). Reconciling Conflicting Results Derived from Experimental and Survey Studies of Altitude Change. *The American Psychologist* **14**, 8–17

Kuhn, T. S. (1962). *The Structure of Scientific Revolutions*. University of Chicago Press, Chicago

Lyle, J. and Hoffman, H. R. (1972). Children's Use of Television and Their Media. *In* E. A. Rubinstein, G. A. Comstock and J. P. Murray (eds) *Television and Social Behavior*, Vol. 4, *Television in Day-to-Day Life Problems of Use*, pp. 129–256. US Government Printing Office, Washington D.C.

Marcuse, H. (1964). *One Dimensional Man*. Beacon Press, Boston, Mass.

Meringoff, L. (1980). *Children and Advertising: An Annotated Bibliography*. Children's Advertising Review Unit, Council of Better Business Bureaus, Inc., New York

National Association of Broadcasters (1978). *The Television Code*, 20th edn. The Code Authority of the NAB, New York and Hollywood

Nielsen, A. C. (1975). *The Television Audience, Chicago, Illinois*. A. C. Nielsen Co.

Parsons, Talcott (1949). *The Structure of Social Action*, 2nd edn. Free Press, New York

Prasad, V. K., Rao, T. R. and Sheikh, A. A. (1978). Mother vs. Commercial. *Journal of Communication* **28**, 91–95

Quebec National Assembly (1978). *Consumer Protection Act*, pp. 45–56. L'éditeur officiel du Québec, Quebec City, Quebec, Canada

Reilly Group Inc. (1973). *Assumption by the Child of the Role of Consumer*. Darien, Conn.

Roberts, D. F. and Maccoby, N. (1973). Information Processing and Persuasion: Counter-arguing Behavior. *In* Peter Clark (ed.) *New Models for Mass Communications Research*, Vol. 2. Sage Publications, Sage Annual Reviews of Communication Research, Beverly Hills, California

Roberts, D. F., Bachen, C. M. and Christenson, P. G. (1978). "Perceptions of and Cognitions about Television Commercials and Supplemental Consumer Information". Testimony prepared for the Federal Trade Commission Rulemaking Hearings on Television Advertising and Children. San Francisco, California

Roberts, D. F., Christenson, P. C., Gibson, W., Mooser, L. and Goldberg, M. E. (1980). Developing Discriminating Consumers. *Journal of Communication* 30, 229–231

Robertson, T. S. (1979). Parental Mediation of Television Advertising Effects. *Journal of Communication* 29, 12–25

Robertson, T. S. and Rossiter, J. R. (1974). Children and Commercial Persuasion: An Attribution Theory Analysis. *Journal of Consumer Research* 1, 13–20

Robertson, T. S. and Rossiter, J. R. (1976). Short-Run Advertising Effects on Children: A Field Study. *Journal of Marketing Research* 13, 68–70

Robinson, J. P. and Backman, J. G. (1972). Television Viewing Habits and Aggression. *In* G. A. Comstock and E. A. Rubinstein (eds). *Television and Social Behavior*, Vol. 3, *Television and Adolescent Aggressiveness*, pp. 372–382. US Government Printing Office, Washington D.C.

Ross, R. P., Campbell, T., Huston-Stein, A. and Wright, J. C. (1981a). Nutritional Misinformation of Children: A Developmental and Experimental Analysis of the Effects of Televised Food Commercials. *Journal of Applied Developmental Psychology* 1, 329–347

Ross, R. P., Campbell, T., Wright, J. C., Huston, A. C., Rice, M. L. and Turk, P. (1981b). *When Celebrities Talk, Children Listen: An Experimental Analysis of Children's Responses to TV Ads with Celebrity Endorsement.* Center for Research on the Influence of Television on Children, University of Kansas, Lawrence, Kansas

Rossiter, J. R. (1977). Reliability of a Short Test Measuring Children's Attitudes Towards TV Commercials. *Journal of Consumer Research* 3, 179–184

Rossiter, J. R. (1979). Does TV Advertising Affect Children? *Journal of Advertising Research* 19, 43–49

Rossiter, J. R. (1981). Research on Television Advertising's General Impact on Children: American and Australian Fundings. *In* J. E. Esserman (ed.) *Television Advertising and Children.* Child Research Service, New York

Rossiter, J. R. and Robertson, T. S. (1974). Children's TV Commercials: Testing the Defenses. *Journal of Communication* 24, 137–144

Rossiter, J. R. and Robertson, T. S. (1975). Children's Television Viewing: An Examination of Parent-Child Consensus. *Sociometry* 38, 308–326

Stuart, R. B. and Davis, B. (1972). *Slim Chance in a Fat World: Behavioral Control of Obesity.* Research Press, Champaign, Illinois

Wackman, D. B., Wartella, E. and Ward, S. (1977). Learning to be Consumers: The Role of the Family. *Journal of Communication* 27, 138–151

Ward, S. (1978). Compromise in Commercials for Children. *Harvard Business Review* November–December, 128–136

Ward, S. (1974). Consumer Socialization. *Journal of Consumer Research* 1, 1–14

Ward, S., Wackman, D. and Wartella, E. (1977). *How Children Learn to Buy: The Development of Consumer Information Processing Skills.* Sage Publications, Beverly Hills, California

Wartella, E. (1980). Individual Differences in Children's Responses to Television Advertising. *In* E. L. Palmer and A. Dorr (eds) *Children and the Faces of Television*, pp. 307–322. Academic Press, New York

Factors Influencing the Effect of Television Violence on Children

L. Rowell Huesmann and Leonard D. Eron

Introduction

In the explosion of research and writing about television that has occurred over the past ten years, no topic has aroused more controversy than television violence. Some may have thought the issue was settled ten years ago when the Surgeon General of the United States concluded that there was a "causative relationship between televised violence and subsequent antisocial behavior and the evidence is strong enough that it requires some action on the part of responsible authorities, the TV industry, the Government, the citizens" (Steinfeld, 1972, p. 28). However, the controversy has not subsided, and there is little evidence that significant programing changes have transpired in the United States as a result of the research. Part of the problem may be that commercial television has a vested, profitable interest in violence. Violence attracts viewers, and advertisers pay networks in direct proportion to the number of viewers they attract. It is also true, though, that many observers have yet to be convinced that television violence is harmful to children. One explanation may be that too much emphasis has been placed on the collection of vast amounts of empirical data on television violence and too little attention has been directed at organising these results in a coherent theoretical framework. Another is that those forces with a vested interest in violence have made sure that negative or inconclusive results receive wide publicity. In any case, an examination of the studies undertaken since the Surgeon General's report provides no basis for a change in conclusions. The evidence seems compelling that excessive viewing of television violence is detrimental to many children. Furthermore, it is now somewhat clearer how television exerts its influence.

While many explanations have been offered for the observed relations between television violence and aggression, few have been elaborated formally. Too often researchers have used terms such as observational learning, catharsis or desensitisation very loosely. Instead of developing detailed models of the psychological processes postulated to mediate the television violence–aggression relation, researchers have concentrated on collecting data. The outcome has been a large body of data that is difficult to fit into any comprehensive explanatory model. Few, if any, process theories have been negated, because none have been developed formally enough to be readily falsifiable. The emphasis has been on describing relations between variables rather than on discovering and elucidating processes.

Despite the paucity of hypothesis-testing research exploring specific process models, the existing empirical research does establish several facts. First, there can be no doubt that in most settings there is a positive correlation between the amount of violence a child views on television and how aggressively the child behaves toward others. More aggressive children watch more violent television. This fact had been firmly established even before the Surgeon General's report was released over a decade ago. In that report Chaffee (1972) reviewed the previous literature and concluded that significant positive correlations of 0.15 to 0.30 are the rule. Differences in sampling procedures and techniques for measuring violence viewing or aggressive behavior have substantial effects on the strength of the relation found, but it is always found. To some observers a factor that accounts for only 5% to 10% of the variance in aggressive behavior may not seem important. However, no one has found any variable that can explain much more of the variance in aggressive behavior than that.

The laboratory and field research carried out since the Surgeon General's report have only confirmed more strongly these conclusions of a positive relation (Lefkowitz and Huesmann, 1980; Huesmann, 1982a; Comstock, 1980). But, of course, the controversial question is why such a positive relation obtains. At least four kinds of processes have been postulated that deserve serious consideration: (1) observational learning through which aggressive behaviors depicted on television are learned by a viewer; (2) changes in emotional or physiological arousal and responsiveness that are engendered by violence viewing and affect aggressiveness; (3) attitude changes that result from exposure to television violence and then affect behavior and (4) justification processes in which violence is watched by the aggressive child because it provides an opportunity to rationalise his/her own aggressive behavior as the norm.

Three of these processes, observational learning, attitude change, and justification, clearly predict that a positive relation obtains between violence viewing and aggression. The proposed theories concerning arousal generally are assumed to yield the same prediction, but possess some inherent contradictions that might be used to explain an opposite outcome (Huesmann, 1982a). More importantly, each of these processes also implies that a child's viewing of violent television increases the likelihood that the child will behave aggressively. This does not rule out causality in the opposite direction as well. For example, according to the justification hypothesis, one would expect children who are more aggressive to seek out more violent television in order to find justification for their behavior. This viewing of violence would then make it more likely that the children would behave aggressively again. Too often researchers have viewed the causal relation between violence viewing and aggression as an either/or proposition when, in fact, the casual effect may be bidirectional. Each of these processes also suggests that various mediating variables should play an important role in the violence-aggression relation. In a recent paper Huesmann (1982a) has discussed the evidence for each of these processes. However, in the current chapter we will concentrate on the evidence concerning the direction of the causal relation between violence viewing and aggression and on the role of various mediating variables in the processes.

Does television violence engender aggressive behavior?

There can be little doubt that in specific laboratory settings exposing children to violent behavior on film or television increases the likelihood that they will behave aggressively immediately afterwards. Large numbers of laboratory studies demonstrated this even before the Surgeon General's report appeared (Comstock, 1980). This effect has usually been attributed to "observational learning" (Bandura, 1977; Bandura, Ross and Ross, 1961, 1963a) in which children imitate the behaviors of the models they observe. Just as they learn cognitive and social skills from watching parents, siblings and peers, they learn to behave aggressively from watching violent actors. While the research has illuminated some of the conditions under which behaviors portrayed in the media are most likely to be imitated, the actual importance of observational learning in determining the aggressiveness of children could not be settled by such laboratory experiments. The question is whether the positive correlations between violence viewing and real aggressive behavior found in the field exist because children are imitating the violent behaviors they see on television.

The Lefkowitz, Eron, Walder and Huesmann study (1977; Eron, Huesmann, Lefkowitz and Walder, 1972) provided the first substantial evidence from a field setting implicating television violence as a cause of aggressive behavior. Without rehashing tired arguments, the results suggested that excessive violence viewing increases the likelihood that a child will behave aggressively. While many researchers have appropriate reservations about the analyses used to extract causal inferences from these longitudinal observational data (Kenny, 1972; Comstock, 1978), the critiques advocating a complete rejection of the results (e.g. Armour, 1975; Kaplan, 1972) have contained such serious errors of reasoning (Huesmann, Eron, Lefkowitz and Walder, 1973, 1979) that they have not had a major impact. Since the Lefkowitz et al. (1977) study, a number of other observational studies and field experiments have suggested that violence viewing is a precursor of aggression (Stein and Friedrich, 1972; Leyens, Parke, Camino and Berkowitz, 1975; Parke, Berkowitz, Leyens, West and Sebastian, 1977; Belson, 1978; Huesmann et al., 1979; Singer and Singer, 1981).

In a project funded by the CBS television network, Belson (1978) collected data on 1650 teenage boys in London. Though he did not obtain longitudinal data, on the basis of analysing matched subgroups, he concluded that "the evidence . . . is very strongly supportive of the hypothesis that high exposure to television violence increases the degree to which boys engage in serious violence" (p. 15). More causally conclusive were the data of Singer and Singer (1981). They followed a sample of three- and four-year-olds over the course of a year and carefully measured a number of variables at four different times. A variety of different multivariate analyses of these data all point to the same conclusion – that television viewing, particularly violence viewing, is a cause of heightened aggressiveness in that age children. The Singers' investigation is particularly noteworthy because the researchers distinguished between the different processes by which media violence might affect children and attempted to test the role of a number of cognitive and familial mediators in the relation. The concluded that among their preschool subjects there was an even stronger relation than had been shown with older children in the previously noted investigations. The Singers state, "our data reflecting a third or fourth of the life span of preschoolers seem to point to . . . (a) . . . causal link between watching TV, especially programs with violent content, and subsequent aggression. Certainly our results seem to argue against attributing the later watching of violent TV fare to an aggressive trend in personality or to some third underlying factor" (Singer and Singer, 1981, p. 115). However, Milavsky, Kessler, Stipp and Rubens (1982), on the basis of a longitudinal study of several large samples, reached the conclusion that television violence does not engender aggressive behavior in any children. Funded by NBC, the study was completed several years ago, but the complete results have not yet been published in a scientific form. Reconciliation of Milavsky et al.'s results with the studies that are supportive of a causal connection will have to await such publication.

Several field experiments have been undertaken in which the researchers attempted to control what the child viewed on television. While most of these experiments have had flaws, the majority (Stein and Friedrich, 1972; Leyens et al., 1975; Loye, Gorney and Steele, 1977) have yielded evidence of a positive relation between violence viewing and aggression. For example, in one recent field experiment (Parke et al., 1977), juveniles in institutions in the United States and Belgium were exposed to five days of violent or control films. In both countries, these children who saw the more violent films were observed acting more aggressively during the five days. Two well-known field experiments that found no relation (Feshbach and Singer, 1971; Milgram and Shot-land, 1973) demonstrate the difficulty of generalising the techniques successfully used in a laboratory to a field setting. However, many more plausible explanations exist for their lack of results than that violence viewing and aggressiveness are unrelated (Comstock, 1980).

Two recent studies of the impact of television on previously unexposed populations have also suggested a link to aggressiveness from television viewing. Williams (1978) collected data on a small community in Canada before and after television was introduced in 1973. She compared these data with data collected at the same times from two communities which had had television for many years. The pre-post increases in both verbal and physical aggression by primary school children were significantly greater for the experimental town than for the two control towns. In a similar study Granzberg and Steinbring (1980) compared a Cree Indian community into which television was being introduced with a control Indian community and a control Euro-Canadian community. No pre–post differences in levels of aggression occurred between the experimental and control communities, taken as a whole. But, when children were classified by amount of daily exposure to television significant differences in aggressive attitudes emerged. The introduction to television into the community increased the aggressiveness of those children who watched a lot of television.

THE CHICAGO CIRCLE STUDY

Now, let us consider the results of an ongoing longitudinal study of television violence and aggression being carried out by Huesmann and Eron at the University of Illinois at Chicago Circle and their collaborators in several countries (Huesmann, Eron, Klein, Brice and Fischer, in press a; Bachrach, 1982; Eron, 1980; Eron and Huesmann, 1980; Eron, Huesmann, Brice, Fischer and Mermelstein, 1983; Fraczek, 1980; Huesmann, 1982a; Huesmann, Eron and Lagerspetz, in press b; Lagerspetz and Engblom, 1979; Rosenfeld, Huesmann, Eron and Torney-Purta, 1982; Sheehan, 1982; Baarda, Kuttschreuter and Wiegman, 1982). This longitudinal study involved interviewing and testing a substantial sample of first and third graders, retesting them one year later and again after two years. The samples studied so far have come from the United States (758 children), Australia (289 children), Finland (220 children), Israel (189 children), The Netherlands (469 children) and Poland (237 children). Because we will be discussing the results of this research in the USA throughout this chapter, it is worthwhile to describe the methodology in some detail.

Subjects and methods

The original group of subjects in the USA was comprised of 672 children in a suburban school system near Chicago, Illinois and 86 children from two parochial schools in Chicago. Half of these children were in the first grade at the start of the study in 1977 and half in the third grade; and they were approximately equally divided by sex. In 1978, 607 children remained in the sample and in 1979 there were 505. Almost all of the subject mortality was attributable to children leaving the school systems.

All children were tested in their classrooms in two hourly sessions, a week apart, in 1977, 1978, and 1979. Thus, for the younger cohort, data were obtained in the first, second, and third grades, and for the older cohort, data were obtained in the third, fourth, and fifth grades. Such an overlapping longitudinal design with two cohorts is ideal for studying developmental trends and making causal inferences. In addition to interviewing the children, we interviewed at least one parent of 591 children in 1977 and reinterviewed 304 in 1979.

The primary dependent variable was peer-nominated aggression. Every child in each classroom nominated as many of the other children as he/she wished on 15 descriptive statements. Ten of these deal with aggressive behavior, e.g. "Who pushes and shoves other children?", and a child's aggression score is the proportion of times he/she is nominated on these items. The reliability of this instrument as well as its concurrent, predictive and construct validity have been amply demonstrated (Eron, Walder and Lefkowitz, 1971; Lefkowitz et al., 1977; Walder, Abelson, Eron, Banta and Laulicht, 1961). In the current study coefficient alpha was 0.95, and the test-retest reliability over one months was 0.91.

Television viewing behavior was measured by asking a child to select the "show you watch the most" from a list of ten programs. Eight such lists were presented; so each child selected a total of eight shows. The 80 television shows used on the lists were chosen from the Nielsen data as the most popular for children aged six to eleven. The shows were distributed among the lists so that each contained representatives of violent and nonviolent programs from different times of the week. The violent and nonviolent programs were also equated for sex of central character and popularity. After selecting their favorite program on the list, the child checked an appropriate box indicating whether he/she watched the program "every time it's on; a lot but not always; or, only once in a while". Two independent judges viewed all 80 programs and rated them for the amount of visually portrayed physical aggression on a five-point scale from "not violent" to "very violent". Interrater reliability was 0.75. A child's television violence score was the sum of the judges' mean violence ratings for the child's eight shows weighted by the frequency which the child reported watching the program. Test–retest reliability of this score over one month was also 0.75.

Results

While all of the data have not yet been analysed, some results are available from the United States, Australia, Finalnd, and Poland. As Table 1 reveals, in each of these countries significant positive relations were found between television violence viewing and peer-nominated aggression in most grades. Interestingly, the strength of the relation seems to generally increase with age. In contrast to our previous results, which were significant only for boys (Lefkowitz *et al.*, 1977), positive relations were obtained in the USA for both boys and girls.

Table 1 Correlations between television violence viewing and peer-nominated aggression

Grade	All subjects	Males	Females
USA ($N = 758$)			
1st	0.212^c	0.160^a	0.210^b
2nd	0.234^c	0.204^a	0.245^c
3rd	0.232^c	0.191^a	0.205^a
4th	0.224^c	0.184^a	0.260^c
5th	0.261^c	0.199^a	0.294^c
Finland ($N = 220$)			
1st	0.141	0.026	0.139
2nd	0.163	0.266^a	0.022
3rd	0.257^b	0.064	0.192
4th	0.228^a	0.381^c	-0.158
5th	0.223^a	0.278	0.035
Poland ($N = 237$)			
1st	0.227^b	0.296^b	0.070
2nd	0.186^a	0.170	0.165
3rd	0.293^c	0.259^b	0.236^a
4th	0.230^b	0.200	0.127
Australia ($N = 290$)			
1st	-0.023	-0.081	0.044
3rd	0.223^b	0.171	0.236^a

^{a} $p < 0.05$ ^{b} $p < 0.01$ ^{c} $p < 0.005$

Table 2 Multiple regressions predicting later (third year) aggression from television violence viewing controlling for initial aggression (first year)

Predictor	Standardised regression coefficients	
	Girls ($N = 207$)	Boys ($N = 191$)
Grade	0.267^c	0.132^b
Initial aggression	0.503^c	0.695^c
Average TV violence viewing in first 2 years	0.122^a	0.079
	$r^2 = 0.386$, $F(3,203) = 42.6$, $p < 0.001$	$r^2 = 0.547$, $F(3,187) = 75.3$, $p < 0.001$

^{a} $p < 0.05$ ^{b} $p < 0.01$ ^{c} $p < 0.001$

Table 3 Mean peer-nominated aggression in third year as a function of earlier television viewing and aggression

Initial peer-nominated aggression	Girls		Boys	
	TV violence viewing for first 2 years		Product of TV violence viewing and identification with aggressive TV characters for first 2 years	
	Low	*High*	*Low*	*High*
High	0.284	0.333	0.386	0.401
	(N = 21)	(N = 42)	(N = 46)	(N = 53)
Medium	0.141	0.167	0.164	0.228
	(N = 47)	(N = 43)	(N = 37)	(N = 32)
Low	0.098	0.132	0.088	0.130
	(N = 53)	(N = 37)	(N = 41)	(N = 21)
	0.147	0.214	0.207	0.295
	(N = 121)	(N = 122)	(N = 114)	(N = 106)

A causal analysis of the USA data is shown in Table 2. Rather than utilising cross-lagged correlations, which have several drawbacks (Huesmann, 1982b; Rogosa, 1980), our approach is to present multiple regression analyses (path analyses) indicating the effect of television violence on later aggression controlling for earlier aggression. Table 2 reveals a significant positive effect of television violence viewing on later aggression for girls but only a marginal effect for boys. The means for girls in Table 3 illustrate these results more clearly. No matter what the initial level of aggression of girls, those who watch more television violence are likely to become more aggressive than those that do not. However, the data do not support a conclusion of unidirectional causality. Regressions predicting television violence viewing from aggression revealed that those girls who are more aggressive also are more likely to become heavier violence viewers regardless of their initial level of violence viewing. The bidirectionality of the effect suggests that a simple observational learning model is not sufficient to explain the correlation between violence viewing and aggression for girls. The differing strength of the relations found in the different countries, grades and genders suggests that a number of processes may be operating and that certain variables may be playing an important role in mediating the relation between television violence and aggression. In fact, as will be shown later, a significant causal effect for boys is revealed when the effects of such mediating variables are considered conjointly.

Mediating variables in aggression: television violence relation

FREQUENCY OF VIEWING

One important mediating variable would obviously be the frequency with which a child watches television in general and particularly the frequency with which he/she watches violent programs. A violent program that is viewed only once in a while would not be expected to have as much effect as a violent program viewed regularly. While older studies (Eron *et al.*, 1972; Robinson and Bachman, 1972) had found no relation between total amount of viewing and aggression, McCarthy, Langner, Gersten, Eisenberg and Orzeck (1975) reported that frequency was related to aggression. Similarly the two studies of areas where television was recently introduced that were described earlier in this chapter (Granzberg and Steinbring, 1980; Williams, 1978) both suggested that frequency of viewing was a critical variable. In these and the McCarthy *et al.* (1975) study, amount of television viewed appeared to be the critical potentiating variable in elucidating the relation between violent television and aggressive behavior.

In the current Chicago Circle study the results support this interpretation of frequency as a potentiating variable. The correlations between frequency and peer-nominated aggression were positive and significant. Of course, frequency in this context means the frequency with which the child watches his/her eight favorite programs, not the total number of hours watched per week. As a matter of fact, the child's report of the frequency that he/she watches his/her favorite programs correlated only 0.20 ($N = 540, p < 0.001$) with the mother's report of the number of hours the child watches each week, and mother's report did not correlate with the child's aggressiveness. The correlations in Table 1 between television violence viewing and aggression are based on the violence ratings for the child's favorite shows weighted by the frequency with which the child reported watching them. A multiple regression analysis predicting aggression from the product of the frequency and the violence rating indicated that the strongest relation could be obtained if violence ratings, scaled from 4 for "most violent" to 0 for "nonviolent", were multiplied by frequency, scaled from 10 for watching "every time it's on" to 0 for watching "only once in a while". This was the method used to compute the violence viewing scores in Table 1. In other words, a show that is only viewed once in a while does not have a significant effect on a child's aggressiveness, no matter how violent the show is. In fact, a violence viewing score unweighted for frequency did not correlate at all with aggressiveness. On the other hand, a pure self-reported frequency score correlated as highly as the violence score for some grade-gender combinations. It may be, as some arousal theorists have argued, that excessive viewing, regardless of content, stimulates aggressive behavior.

While it might be tempting to attribute the correlations between self-reported frequency of viewing favorite programs and aggression to methodological or socio-cultural artifacts, such explanations find little support in the data. Response bias due to impulsivity, for example, was measured (Rosenfeld, Huesmann, Eron and Torney-Purta, 1982) and does not explain the relation. Similarly, controlling for social class did not reduce the television-aggression correlation significantly, though for boys lower social class was correlated both with more violence viewing and greater aggressiveness.

It is also true that television violence viewing relates to a child's popularity with his or her peers. In our ten-year longitudinal study we had found that the less popular child turns to watching more and more television. At the same time, the less popular child tends to be a more aggressive child. In the current cross-national study we found significant negative correlations between popularity in all countries and grades and both sexes. One may hypothesise that the less popular child, lacking social reinforcements, watches television to obtain vicariously the gratifications denied in social interactions.

INTELLIGENCE OF THE CHILD

Intelligence has often been postulated as a third variable that might explain the relation between television violence viewing and aggression; however, the evidence has not generally supported such a thesis. In most observational studies a significant relation between violence viewing and aggression remains when the effect of intelligence has been partialed out (Belson, 1978; Lefkowitz et al., 1977; Singer and Singer, 1981). In our ten-year longitudinal study, evidence was uncovered that suggested that low achievement led a child to watch more and more television. Since low achievement and low IQ can be frustrators that may engender aggression, it is not surprising that low achievement and low IQ were also correlated with aggression in that study. However, these correlations could not explain away the relation between violence viewing and aggression. In the current Chicago Circle study low achievement is again one of the higher correlates of aggressive behavior ($r = -0.38$, $p < 0.001$ for a two-year lag). Measured with a standardised test administered by the school system, these reading achievement scores also correlated significantly negatively with frequency of viewing favorite programs, with the violence ratings of the favorite programs, and with the composite violence viewing score. In the current study, when achievement was partialed out, there was no significant change in the relation between television violence viewing and aggression for girls. However, for boys the partial correlation between violence viewing and aggression, controlling for reading achievement, was substantially lower than the raw correlation.

Boys who are lower achievers watch their favorite programs more regularly, prefer more violent programs and behave more aggressively. A number of different hypotheses would be consistent with this result, but they are very difficult to distinguish on the basis of observational data.

Perhaps the most plausible hypothesis is that the child reacts to poor achievement by turning to television violence for vicarious gratification, and this increased involvement in television in turn hampers school achievement. At the same time, the frustration of low achievement coupled with the exposure to television violence leads the child to more aggressive behaviors.

AGE OF THE VIEWER

A number of researchers have attempted to determine the ages at which children are most likely to imitate observed behaviors. Eron *et al.* (1972) argued that once an individual has reached adolescence, behavioral predispositions and inhibitory controls have become crystallised to the extent that a child's aggressive habits would be difficult to change with modelling. Collins (1973; 1981; Collins, Berndt and Hess, 1974; Newcomb and Collins, 1979) has consistently found that young children are less able to draw the relation between motives and aggression and therefore may be more prone to imitate inappropriate aggressive behaviors. Hearold's (1979) review of the area generally supports these views, but suggests that modelling might increase again among adolescent boys. Perhaps the more important question, however, is at how young an age children begin to imitate behaviors viewed on television. McCall, Parke and Kavanaugh's (1977) experiments indicate that children as young as two years were facile at imitating televised behaviors, and some imitation was observed in even younger children. The Singer and Singer (1981) study described earlier provided evidence of adverse effects for television violence on children who were only three- or four-years-old.

In the current cross-national study we have found evidence of a relation between television violence and aggression in all ages from six to eleven. However, Table 1 suggests that the relation may become stronger as children grow older. Part of this trend may be due to a methodological factor – the measures used are somewhat less reliable with first graders than with older children. However, it is also true, as Fig. 1 reveals, that television violence viewing seems to increase dramatically from the first to the third or fourth grade. Correspondingly, over the same period of time, the child's perception of television violence as realistic is decreasing precipitously, as Fig. 2 indicates. Thus the third grade may be the center of a critical period when the factors are rife for television violence to have an effect. Interestingly, in this regard some of the strongest relations between television violence and both simultaneous and later aggression have been reported for children about this age (Chaffee, 1972; Lefkowitz *et al.*, 1977).

GENDER OF VIEWER AND OBSERVED CHARACTER

An important finding in early field studies of aggression and television violence was that females were less affected by violence viewing than males (Bailyn, 1959; Eron, 1963). In our ten-year-old longitudinal study,

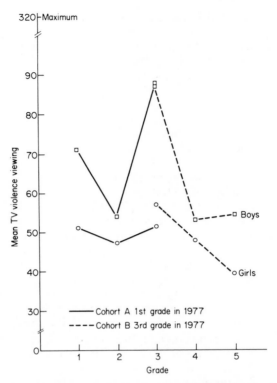

Fig. 1   Television violence viewing as a function of a child's age in two overlapping cohorts

conducted between 1960 and 1970, we found no correlation between a girl's violence viewing and her later aggressiveness (Lefkowitz *et al.*, 1977). However, that situation no longer seems to hold. In our cross-national study we have found positive correlations between violence viewing and aggression for girls in at least some grades in all countries. In the Chicago area sample, though, the correlations were even higher for girls than for boys.

One hypothesis that might explain these results would be that in the 1960s there were very few aggressive females on television for girls to imitate, but now there are. The lower correlations for girls in other countries where there are very few aggressive female models on television would be consistent with such a theory. According to observational learning theory, imitation would be greater if the viewer can identify with the model. Within the existing literature, however, the evidence is ambiguous on the role that gender identification plays in observational learning. Bandura, Ross and Ross (1963a,b) found that both boys and girls more readily imitated male rather than female models. To test the gender identification hypothesis in our Chicago Circle study, we scored

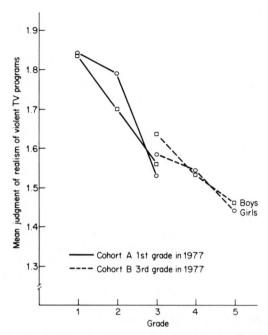

Fig. 2 Mean judged realism of television violence as a function of a child's age in two overlapping cohorts

television programs for the amount of violence perpetrated by males and females. We found that regardless of the child's gender there were higher correlations between the child aggressiveness and the child's viewing of a male character's violence than between aggressiveness and the child's viewing of a female character's violence. This apparently greater influence of male models on children has been detected in our data from other countries as well. Thus, it does not seem reasonable to attribute the emergence of a relation between violence viewing and aggression in girls to the appearance of female models on television.

One of the problems with using gender as a measure of identification with a television model is that aggression is highly correlated with sex-role orientation (Eron and Huesmann, 1980; Huesmann *et al.*, 1978; Lefkowitz *et al.*, 1977). Girls who are aggressive may in fact identify more with male actors than with most female actors. It may be that more important than the sex of the model are the behaviors the model is performing and that, if masculine activities are intrinsically more appealing to subjects of either sex, then all subjects would be more likely to attend to male characters and be influenced most by their behaviors. Along these lines, it has been demonstrated that the more powerful the model, the more likely are the model's behaviors to be copied regardless of sex (Bandura, Ross and Ross, 1963c).

In our current cross-national study we measured sex-role orientation by having children select the activities they preferred most from six pictures representing two stereotype males, two stereotype females and two neutral activities. They repeated this task for four sets of pictures, and they received three scores representing the number of masculine, feminine, and neutral activities chosen. The pictures had been selected for use on the basis of 67 college students' ratings of their stereotypicality, and the measure was demonstrated to have good reliability. Table 4 shows the correlations between neutral sex-role orientation and peer-nominated aggression for boys and girls over the course of our three-year study. While the relations between aggression and either a male or female orientation varied greatly with sex and grade, the relation between aggression and neutral orientation was consistently negative. Children who scored high on neutral sex-role were ones who were flexible in their choice of games and activities and not bound by societal stereotypes. Perhaps such children are also more flexible in their choice of behaviors in frustrating situations and therefore less aggressive.

While for boys the preference for masculine activities remained relatively constant with age, for girls it increased significantly with age ($r = 0.284$, $p < 0.001$). At the same time, as Table 1 reveals, the relation between television violence and aggressiveness was increasing somewhat with age for girls. Also, masculinity in the girls was correlated significantly with aggressiveness which suggests that it is the girls who are behaving more like boys who are responsible for the relation between violence viewing and aggressiveness. The hypothesis is that the recent greater emphasis on the need and desirability for females to be assertive and physically active has resulted in girls having fewer inhibitions about modelling the aggressive behaviors they observe on television. Thus, many girls and especially those who prefer the traditional active, competitive and violent masculine activities, use the actors they observe on television as models for their behavior. But why should the correlations between television violence viewing and aggressive behavior now be

Table 4   Correlations between preference for neutral sex-typed activities and peer-nominated aggression

| Grade | Correlations | |
	Girls	Boys
1st	−0.197[a]	−0.175[a]
2nd	−0.294[b]	−0.086
3rd*	−0.135	−0.151
4th	−0.149	−0.200[a]
5th	−0.140	−0.117

[a] $p < 0.05$ [b] $p < 0.01$
* The correlations for the two cohorts have been averaged for the 3rd grade.

even higher for girls than for boys? There are two possible reasons. From an observational learning perspective, these girls have available both male and female aggressive models to copy while boys are probably inhibited in imitating females. More generally, girls have a much lower average level of aggressiveness and are exposed less often to aggressive models in their everyday interactions. Thus, there is a greater potential for television violence to change their behavior.

IDENTIFICATION WITH TELEVISION CHARACTERS

While the weight of evidence seems to indicate that all viewers are most likely to imitate an heroic, white male actor, individual differences should not be ignored. It may be that some children identify much more with some actors and this identification mediates the relation between violence viewing and aggressiveness. Such an identification would be important not just in an observational learning model but also in a justification or attitude change model. The more the child identifies with the actors who are aggressors or victims, the more likely is the child to be influenced by the scene.

To test this theory in our Chicago Circle study, we asked each child how much he/she was like several characters on television. The characters included two aggressive males, two aggressive females, two unaggressive males, and two unaggressive females. From their responses a reliable "identification with aggressive character" score was derived.

Not surprisingly, the score correlated significantly with aggressiveness, particularly for boys. More interesting was the discovery that identification with aggressive television characters interacted with violence viewing to establish an even stronger relation with aggression. Table 5 shows a multiple regression predicting later aggression from the product of violence viewing and identification with aggressive characters after earlier aggression was partialed out. The effect is significant for boys. Those boys who watch violent television and identify with aggressive television characters are predictably more aggressive two years later regardless of their initial level of aggressiveness. This is perhaps the most important finding in the current longitudinal study – that identification with aggressive characters seems to be a catalyst substantially increasing the effect of television violence on boys. Identification with aggressive television characters by itself is a good predictor of aggression, but not as significant a predictor as its product with television violence viewing.

The relation is illustrated by the means for boys in Table 3. Regardless of his initial level of aggressiveness, the boy who watches more violence and identifies with the violent character, is more likely to increase in aggression over three years. Furthermore, as Table 5 illustrates, this effect is mostly unidirectional from the television variables to aggression. Aggression is not nearly as good a predictor of the television variables as the television variables are of aggression.

It is interesting that this effect only obtains for boys. For girls there is a

Table 5 Multiple regressions relating boys aggression to a multiplicative composite of television violence viewing and identification with aggressive television characters controlling for initial level of the dependent variable ($N = 191$)

Dependent variable	Predictors	Standardised regression coefficients
Aggression in third year		
	Grade	0.151^b
	Aggression in first year	0.679^c
	Product of TV violence viewing and identification with aggressive characters for first 2 years	0.152^b
		$r^2 = 0.575, F(3,187) = 80.3, p < 0.001$
Product of TV violence viewing and identification with aggressive TV characters in third year		
	Grade	-0.236^c
	Product of TV violence viewing and identification with aggressive characters in first year	0.384^c
	Aggression in first 2 years	0.102
		$r^2 = 0.232, F(3,187) = 18.9, p < 0.001$

$^a\ p < 0.05$ $^b\ p < 0.01$ $^c\ p < 0.001$

slight correlation between aggressive behavior and identification with aggressive television characters, but the latter variable does not add to the power of the violence viewing variable in predicting a girl's aggression. However, this result is not anomalous. In the only other country in which sufficient data are available for a longitudinal analysis at this time – Finland, the identical results were obtained! While television violence alone was not a very good predictor of a boy's later aggression, the product of violence viewing and identification with aggressive characters was a highly significant predictor regardless of the child's initial level of aggression. But, as in the USA, earlier aggression was not a significant predictor of later scores on this composite television viewing variable. Thus, identification with television characters seems to be a significant mediating variable in the relation between television violence and aggressive behavior in boys.

FANTASY–REALITY DISCRIMINATIONS

Another potential mediating variable in the relation between television violence and aggressive behavior might be a child's ability to discriminate between fantasy and reality as portrayed on television. Violence scenes perceived as unrealistic by the child would be less likely to affect the child's behavior under any of the process models. Some evidence for such an effect has been provided by Feshbach (1976). As a result, one might expect that individual differences on this variable could be very important in determining who would be most affected by television violence.

In the current Chicago Circle study, we measured children's perception of television violence as fantasy or reality by asking them "how much do you think 'program X' tells about life like it really is? – Just like it really is? – A little like it is? – Not at all like it is?" They were asked this question about ten violent programs, and their scores were the sum of their responses for the shows they had watched.

In the earlier ten-year longitudinal study (Lefkowitz et al., 1977) it had been found that girls thought television was significantly less realistic than boys. It was hypothesised that this might be one of the reasons for the lack of a significant longitudinal relation between violence viewing and aggression for girls at that time. In support of this hypothesis were data indicating that the more aggressive a girl was at both age eight and age nineteen, the more realistic she thought television was. In the current Chicago sample, however, we found that girls and boys now perceive television violence to be equally realistic. This adds validity to the theory that fantasy–reality discriminations also mediate the effect of television violence on aggression since, as pointed out above, girls and boys now display an equally strong relation between violence viewing and aggression.

In our current data we have found positive relations between aggression and this realism variable for both sexes. These positive correlations ranged from 0.11 to 0.25 depending on the gender and grade of the subjects with no systematic differences (though the child's perception of television violence as realistic declines dramatically with age).

AGGRESSIVE FANTASY

One final mediating variable that should be considered is the child's use of aggressive fantasy. Some theorists have argued that a child who reacts to television violence by fantasising about aggressive acts might actually become less aggressive (Feshbach, 1964). While no researcher has ever reported finding such negative correlation in a field study, this variation of a catharsis theory still raises its head from time to time. A more compelling argument exists that fantasising about aggressive acts should lead to greater aggression by the child. From an information processing perspective the rehearsal of specific aggressive acts observed on television through daydreaming or imaginative play should increase the probability that the aggressive acts will be performed. In support of this theory, Singer and Singer (1981) report that children who engage in more prosocial imaginative play and fantasy are less aggressive. The hypothesis is that these children have rehearsed prosocial behaviors sufficiently for them to become dominant responses.

In the current cross-national study aggressive and active-heroic fantasy were measured with the Childrens Fantasy Inventory (Rosenfeld *et al.*, 1982). On this 45-item questionnaire children report how often they engage in different types of fantasy activity. In this Chicago sample we found significant positive correlations between peer-nominated aggression and both fantasy variables for boys and girls. The correlation between aggression and active-heroic fantasy was highest for girls ($r = 0.17$, $p < 0.001$) and the correlation between aggression and aggressive fantasy was highest for boys ($r = 0.20$, $p < 0.001$). These results are in agreement with the hypothesis that aggressive fantasising serves as a cognitive rehearsal of aggressive acts and increases the likelihood of their emission.

Mitigating the effects of television violence

One of the best ways to demonstrate the importance of mediating variables in the relation between television violence and aggression is by intervening to manipulate those variables. Since the use of procedures which might enhance the effect of violence viewing on children is ethically questionable, the appropriate technique is to devise methods aimed at mitigating the effect. A few investigators working with preschool children have attempted to do this. Friedrich-Cofer, Huston-Stein, Kipnis, Susman and Clevit (1979) demonstrated that the effects of prosocial television was greatly enhanced when it was coupled with other prosocial teaching. Hicks (1965) discovered that adults' comments about an aggressive scene influenced the likelihood that a preschooler would imitate the scene so long as the adult was present, while Singer and Singer (1981) reported that a parent's presence, by itself, had no effect. On the other hand, Grusec (1973) found that with older children an adult's comments could have lasting influence.

In the current Chicago Circle study (Huesmann *et al.*, in press a, b) 132 children, aged six to ten, who had already been identified as high violence viewers were randomly assigned to one of two groups. The first group (the experimental group) was exposed to two series of interventions designed to convince children that television violence was unrealistic and should not be imitated. The other group (the placebo group) received a control treatment. The first series done at the beginning of the second year of the longitudinal study included three sessions in small groups during which an attempt was made to teach the subjects how unrealistic television violence is. The subjects in the experimental group were shown excerpts from violent shows and listened to a highly structured discussion of how unrealistic the actors' behaviors were and how their problems could have been solved unaggressively. The placebo group was shown nonviolent educational excerpts followed by discussion of their content. Then, at the beginning of the third year, a more formal attitude change procedure was used with the experimental subjects in which the subjects themselves took a more active role. Each of the experimental subjects was asked to write a paragraph on "why television violence is unrealistic and why viewing too much of it is bad". Over the course of two sessions, the children in the experimental group wrote the paragraph, received suggestions and rewrote it, were taped reading the paragraph, and watched a videotape of themselves and their classmates reading the paragraphs. The subjects were told that the tapes were going to be shown to the school children in Chicago. The placebo group also made a tape, but it was about "what you did last summer". Four months after this intervention, the final wave of data on all the children in the study was collected. Remarkably, as shown in Table 6, it was found that the mean peer-nominated aggression score for the experimental group was now significantly lower than the score for the placebo group, although a year previously the two groups were approximately equal in score. The difference was highly significant as evaluated by analysis of covariance with sex, grade and pretreatment aggression score as covariates.

Two analyses suggest what happened in the intervention to bring about the lower aggression scores on the part of high violence viewers who were in the experimental group. Table 7 represents the predictors to aggression from a set of variables including sex, grade, television

Table 6 The effect of the intervention on mean level of aggression over the course of one year

		Mean peer-nominated aggression (peer-agg.)	
		Before (1978)	After (1979)
Experimental group ($N = 59$)		0.154	0.175
Placebo group ($N = 58$)		0.158	0.242
Analysis of covariance	Source	df	F	Significance
Covariates				
	Sex	1	1.23	—
	Grade	1	0.00	—
	Peer-agg., 1978	1	61.12	0.001
Effects				
	Group	1	6.40	0.013
Error		112		
Total		116		

Table 7 Multiple regressions predicting 1979 peer-nominated aggression for experimental and placebo group from grade, sex and 1979 television viewing behavior*

Predictor Variable	Experimental group Raw Corr	Beta	Placebo group Raw Corr	Beta
Sex	-0.268^b	0.394^c	0.260^b	0.231
Grade	-0.042	0.109	0.282^b	0.303^a
TV violence viewing	0.066	0.004	0.281^b	0.192^a
Identification with TV characters	0.385^d	0.550^d	0.129	0.100
Judgment of TV realism	-0.016	-0.292^a	0.153	0.060
	$r^2 = 0.318$, $F(5,44) = 4.10$, $p < 0.006$		$r^2 = 0.222$, $F(5,45) = 2.56$, $p < 0.050$	

[a] $p < 0.05$ [b] $p < 0.025$ [c] $p < 0.01$ [d] $p < 0.005$
* Pairwise deletions of missing data were used since listwise deletions did not yield any major differences.

violence, extent of identification with television characters and judgment of television realism.

The most striking result in this table is the lack of relation between television violence and aggression in the experimental group and the continued positive relation in the placebo group, almost the same degree of relation as in the general population. The best predictor, however, in the experimental group was identification with aggressive television characters. Those subjects who had higher self-rated identification with television characters had higher peer-nominated aggression scores. There was no significant relation in the placebo group between identification and aggression. The sheer amount of violence viewed by that group was sufficient to predict aggression, regardless of whether the subjects identified with the character. Gender, as would be expected, continued to be a good predictor of aggression in both groups – boys were more aggressive than girls in all conditions. Grade level was a predictor only in the placebo group. This is not surprising since grade and aggression, as stated before, are positively related, and, in the placebo group, there was no intervention introduced in order to diminish aggression. In the experimental group the relation disappeared. Actually, the intervention was more successful with the older children in the experimental group. Older children improved considerably more in their attitude toward television than younger children. Furthermore, older children tended to identify less with television characters than did younger children even before the intervention and, after the intervention, decreased in identification score even more. Additionally, as noted above, the intervention apparently did not work for those subjects who identified with television characters. They continued to get higher aggression scores.

Why were those subjects who identified with television characters less susceptible to the treatment? The treatment attempted to change attitudes about television and about aggression as well. As a manipulation check, six attitudinal questions were asked before and after treatment, e.g. "How much of what kids see on television is fake? – How likely is it that watching a lot of violent television shows would make a kid meaner? – Are television shows with a lot of hitting and shooting harmless for kids?" The subjects responded to these questions on a five point scale. The score was the sum of the weighted responses. A change score was calculated for each child by subtracting the pretreatment score from the posttreatment score. The larger the score, the more the child changed toward a more realistic attitude. We found that the most important predictor of change was identification with television characters. The more identified were the subjects with television characters, the less likely were they to change their attitudes toward television as a result of the intervention. It will be remembered also that the more the youngsters identified with television characters the more aggressive they were. Extent of identification with television characters is thus demonstrated to be an important mediating variable in the relation between television violence and aggression. The only other significant predictor was a self-report on the extent to which the subjects read fairy tales or had fairy tales read to them. The more extensive the reading of fairy tales, the more likely was the attitude toward television to change. This latter finding is in keeping with the Singers' contention (1981) that training in fantasy can affect the relation between television violence and behavior.

Summary and conclusions

In this chapter we have attempted to report the latest findings on the relation between television violence viewing and aggressive behavior in children and to determine how these findings contribute to a better understanding of the intervening processes in that relation. There can no longer be any question that there is a positive relation between television violence and children's behavior. Furthermore, there can be little doubt that excessive exposure to television violence increases a child's aggressiveness under certain conditions. However, there has as yet been no unequivocal demonstration of how this relation is caused, what are the essential precursor and intervening variables, or whether the relation is unidirectional or bidirectional. A number of theoretical explanations have been offered, including observational learning, attitude change, arousal and justification processes.

It turns out that these processes are not mutually exclusive. In fact, they may all be contributing in varying degree to the strength of the relation between televised violence and aggression. Nor can any of these processes claim primacy in determining the direction of the relation. Indeed, as we have seen, it is likely both that violent television makes the observing children aggressive and that aggressive children choose to watch violent television. The question of which comes first, television violence or child aggression, is a false question; the process is most likely circular. Aggressive children, as we have seen, are unpopular and because their relations with their peers tend to be unsatisfying they spend more time watching television than their more popular peers; the violence they see on television reassures them that their own behavior is appropriate and at the same time teaches them new coercive techniques which they then attempt to use in their interaction with others, which in turn makes them more unpopular and drives them back to the television and the circle continues. Similarly, aggression is associated with low achievement – for reasons we do not yet know. But we do know that children who do not succeed in school spend more time watching television than scholastically successful children and therefore having more opportunity to observe aggressive models. They tend to identify with these models more than their achieving peers and are therefore more influenced by the violent acts they observe their favorite characters perform. Since their resources for problem solving are more limited than those of scholastically successful youngsters, the easy solutions they observe on television are more readily employed in their interactions with others. This type of behavior isolates them from their peers, leads them to more television with less time for study and so on and so on.

We have demonstrated, however, that it is possible to break into this cycle with simple interventions and short circuit the connection between television violence and aggression. Unfortunately, with the procedures we have used there is less chance for successful intervention when children identify strongly with television characters and believe that the violence they observe on television is an accurate description of "real-life". The simple instructional procedures we have used may have to be modified for children already severely entrenched, by reason of constitution or environment, in these attitudes and beliefs.

References

Armour, D. J. (1975). "Measuring the Effects of Television on Aggressive Behavior". RAND Report #R-1759-MF

Baarda, B., Kuttschreuter, M. and Wiegman, O. (1982). "A Cross-cultural Contribution of the Netherlands to the Research on Aggression and Television Viewing and an Extension to Prosocial Behavior". Paper presented at meetings of *International Society for Research on Aggression*, Mexico City

Bachrach, R. S. (1982). "Aggression and Television Violence Viewing in Kibbutz and City Children". Paper presented at meetings of *International Society for Research on Aggression*, Mexico City

Bachrach, R. S. (1981). Personal communication

Bailyn, L. (1959). Mass Media and Children: A study of exposure habits and cognitive effects. *Psychological Monographs* **73**(1), 1–48

Bandura, A. (1977). *Social Learning Theory*. Prentice Hall, Englewood Cliffs

Bandura, A., Ross, D. and Ross, S. A. (1961). Transmission of aggression through imitation of aggressive models. *Journal of Abnormal and Social Psychology* **63**, 575–582

Bandura, A., Ross, D. and Ross, S. A. (1963a). Imitation of film-mediated aggressive models. *Journal of Abnormal and Social Psychology* **66**, 3–11

Bandura, A., Ross, D. and Ross, S. A. (1963b). Vicarious reinforcement and imitative learning. *Journal of Abnormal and Social Psychology* **67**, 601–607

Bandura, A., Ross, D. and Ross, S. A. (1963c). A comparative test of the status envy, social power and secondary reinforcement theories of identification learning. *Journal of Abnormal and Social Psychology* **67**, 527–534

Belson, W. (1978). *Television Violence and the Adolescent Boy*. Saxon House, Hampshire, UK

Chaffee, S. H. (1972). Television and adolescent aggressiveness (overview). *In* G. A. Comstock and E. A. Rubinstein (eds) *Television and Social Behavior*, Vol. 3, *Television and Adolescent Aggressiveness*. US Government Printing Office, Washington, D.C.

Collins, W. A. (1973). Effect of temporal separation between motivation, aggression, and consequences: A developmental study. *Developmental Psychology* **8**, 215–221

Collins, W. A. (1982). Cognitive processing aspects of television viewing. *In* D. Pearl, L. Bouthilet and J. Lazar (eds) *Television and Behavior: Ten Years of Scientific Progress and Implications for the 80's*. US Government Printing Office, Washington, D.C.

Collins, W. A., Berndt, T. J. and Hess, V. L. (1974). Observational learning of motives and consequences for television aggression: A developmental study. *Child Development* **45**, 799–802

Comstock, G. (1978). A contribution beyond controversy. (Review of *Growing up to be Violent: A Longitudinal Study of the Development of Aggression* by M. M. Lefkowitz, L. D. Eron, L. O. Walder and L. R. Huesmann, Pergamon Press, New York). *Contemporary Psychology* **23**(11), 807–809

Comstock, G. (1980). New emphases in research on the effects of television and film violence. *In* Palmer, E. L. and Dorr, A. (eds) *Children and the Faces of Television: Teaching, Violence, Selling*. Academic Press, New York

Eron, L. D. (1963). Relationship of TV viewing habits and aggressive behavior in children. *Journal of Abnormal and Social Psychology* **67**, 193–196

Eron, L. D. (1980). Prescription for reduction of aggression. *American Psychologist* **35**, 244–252

Eron, L. D. and Huesmann, L. R. (1980). Adolescent aggression and television. *Annals of the New York Academy of Sciences* **347**, 319–331

Eron, L. D., Walder, L. O. and Lefkowitz, M. M. (1971). *Learning of Aggression in Children*. Little Brown, Boston

Eron, L. D., Huesmann, L. R., Lefkowitz, M. M. and Walder, L. O. (1972). Does television violence cause aggression? *American Psychologist* **27**, 253–263

Eron, L. D., Huesmann, L. R., Brice, P., Fischer, P. and Mermelstein, R. (1983). Age trends in the development of aggression, sex typing and related television viewing habits. *Developmental Psychology* **19**, 71–77

Feshbach, S. (1964). The function of aggression and the regulation of aggressive drive. *Psychological Review* **71**, 252–272

Feshbach, S. (1976). The role of fantasy in the response to television. *Journal of Social Issues* **32**(4), 71–85

Feshbach, S. and Singer, R. D. (1971). *Television and Aggression: An Experimental Field Study*. Jossey-Bass, San Francisco

Fraczek, A. (1980). "Cross Cultural Study of Media Violence and Aggression among Children. Comments on Assumptions and Methodology". Paper read at the XXIInd International Congress of Psychology. Leipzig, GDR

Friedrich-Cofer, L. K., Huston-Stein, A., Kipnis, D., Susman, E. J. and Clevit, A. S. (1979). Environmental enhancement of prosocial television content: Effects of interpersonal behavior, imaginative play, and self-regulation in a natural setting. *Developmental Psychology* **15**(6), 637–646

Granzberg, G. and Steinbring, J. (1980). "Television and the Canadian Indian". Technical Report, Department of Anthropology, University of Winnipeg, Winnipeg, Manitoba

Grusec, J. E. (1973). Effects of co-observer evaluation on imitation: A developmental study. *Developmental Psychology* **8**, 141

Hearold, S. L. (1979). "Meta-Analysis of the Effects of Television on Social Behavior". Unpublished doctoral dissertation, University of Colorado

Hicks, D. J. (1965). Imitation and retention of film-mediated aggressive peer and adult models. *Journal of Personality and Social Psychology* **2**, 97–100

Huesmann, L. R. (1982a). Television violence and aggressive behavior. In D. Pearl, L. Bouthilet and J. Lazar (eds) *Television and Behavior: Ten Years of Scientific Progress and Implications for the 80's.* US Government Printing Office, Washington, D.C.

Huesmann, L. R. (1982b). Process models of social behavior. In N. Hirschberg (ed.) *Multivariate Methods in the Social Sciences: Applications.* Erlbaum, Hillsdale, N.J.

Huesmann, L. R., Eron, L. D., Lefkowitz, M. M. and Walder, L. O. (1973). Television violence and aggression: the causal effect remains. *American Psychologist* **28**, 617–620

Huesmann, L. R., Fischer, P. F., Eron, L. D., Mermelstein, R., Kaplan-Shain, E. and Morikawa, S. (1978). "Children's Sex-role Preference, Sex of Television Model, and Imitation of Aggressive Behaviors". Paper presented at the meeting of the International Society for Research on Aggression, Washington, D.C.

Huesmann, L. R., Eron, L. D., Lefkowitz, M. M. and Walder, L. O. (1979). "Causal Analyses of Longitudinal Data: An Application to the Study of Television Violence and Aggression". Technical Report. Department of Psychology, University of Illinois at Chicago Circle, Chicago, Illinois

Huesmann, L. R., Lagerspetz, K. M. and Eron, L. D. Intervening variables in the television violence aggression relation: Evidence from two countries. *Developmental Psychology* (In press a)

Huesmann, L. R., Eron, L. D., Klein, R., Brice, P. and Fischer, P. Mitigating the imitation of aggressive behaviors by changing children's attitudes about media violence. *Journal of Personality and Social Psychology: Attitudes and Social Cognition* (In press b)

Kaplan, R. M. (1972). On television as a cause of aggression. *American Psychologist* **27**, 968–969

Kenny, D. A. (1972). Threats to the internal validity of cross-lagged panel inference, as related to "Television violence and child aggression: A follow-up study". In G. A. Comstock and E. A. Rubinstein (eds) *Television and Social Behavior*, Vol. 3, *Television and Adolescent Aggressiveness.* US Government Printing Office, Washington, D.C.

Lagerspetz, K. M. J. and Engblom, P. M. (1979). Immediate reactions to TV-violence by Finnish pre-school children of different personality types. *Scandinavian Journal of Psychology* **20**, 43–53

Lefkowitz, M. M. and Huesmann, L. R. (1980). Concomitants of television violence viewing in children. In Palmer, E. L. and Dorr, A. (eds) *Children and the Faces of Television: Teaching, Violence, Selling.* Academic Press, New York

Lefkowitz, M. M., Eron, L. D., Walder, L. O. and Huesmann, L. R. (1977). *Growing up to be Violent: A Longitudinal Study of the Development of Aggression.* Pergamon, New York

Leyens, J. P., Parke, R. D., Camino, L. and Berkowitz, L. (1975). Effects of movie violence on aggression in a field setting as a function of group dominance and cohesion. *Journal of Personality and Social Psychology* **32**, 246–360

Loye, D., Gorney, R. and Steele, G. (1977). An experimental field study. *Journal of Communication* **27**(3), 206–216

McCall, R. B., Parke, R. D. and Kavanaugh, R. D. (1977). Imitation of live and televised models by children one to three years of age. *Monographs of the Society for Research in Child Development* **42**(5), 94 pp.

McCarthy, E. D., Langner, T. S., Gersten, J. C., Eisenberg, J. G. and Orzeck, L. (1975). Violence and behavior disorders. *Journal of Communication* **25**(4), 71–85

Milavsky, J. R., Kessler, R., Stipp, H. and Rubens, W. S. (1982). *In* D. Pearl, L. Bouthilet and J. Lazar (eds) *Television and Behavior: Ten Years of Scientific Progress and Implications for the 80's*. US Government Printing Office, Washington, D.C.

Milgram, S., and Shotland, R. L. (1973). *Television and Antisocial Behavior: Field Experiments*. Academic Press, New York

Newcomb, A. F. and Collins, W. A. (1979). Children's comprehension of family role portrayals in televised dramas: Effects of socioeconomic status, ethnicity, and age. *Development Psychology* **15**(4), 417–423

Parke, R. D., Berkowitz, L., Leyens, J. P., West, S. and Sebastian, R. J. (1977). Some effects of violent and nonviolent movies on the behavior of juvenile delinquents. *In* Berkowitz, L. (ed.) *Advances in Experimental Social Psychology*, Vol. 10, pp. 135–172. Academic Press, New York

Rogosa, D. (1980). A critique of cross-lagged correlation. *Psychological Bulletin* **88**, 245–258

Robinson, J. P. and Bachman, J. G. (1972). Television viewing habits and aggression. *In* G. A. Comstock and E. A. Rubinstein (eds) *Television and Social Behavior*, Vol. 3, *Television and Adolescent Aggressiveness*. US Government Printing Office, Washington, D.C.

Rosenfeld, E., Huesmann, L. R., Eron, L. D. and Torney-Purta, J. V. (1982). Measuring patterns of fantasy behavior in children. *Journal of Personality and Social Psychology* **42**, 347–366

Rosenfeld, E., Maloney, S., Huesmann, L. R., Eron, L. D., Fischer, P. F., Musonis, W. and Washington, A. (1978). "The Effect of Fantasy Behaviors and Fantasy-Reality Discriminations upon the Observational Learning of Aggression". Paper presented at the meeting of the International Society of Research on Aggression, Washington, D.C.

Sheehan P. W. (1982). "Television and its impact: A longitudinal study". Report prepared for the *Australian Research Council*. University of Queensland, St. Lucia, Australia

Singer, J. L. and Singer, D. G. (1981). *Television, Imagination and Aggression: A Study of Preschoolers Play*. Erlbaum, Hillsdale, N.J.

Stein, A. H. and Friedrich, L. K. (1972). Television content and young children's behavior. *In* J. P. Murray, E. A. Rubinstein and G. A. Comstock (eds) *Television and Social Behavior*, Vol. 2, *Television and Social Learning*. US Government Printing Office, Washington, D.C.

Steinfeld, J. (1972). "Hearings before the Subcommittee on Communications of the Committee on Commerce United States Senate, Ninety-Second Congress, Second Session". US Government Printing Office, Washington, D.C.

Walder, L. O., Abelson, R., Eron, L. D., Banta, T. J. and Laulicht, J. H. (1961). Development of a peer-rating measure of aggression. *Psychological Reports* **9**, 497–556 (monograph supplement 4–19)

Williams, T. M. (1978). "Differential Impact of TV on Children: A Natural Experiment in Communities with and without TV". Paper presented at the meeting of the International Society for Research on Aggression, Washington, D.C.

The Effects of Televised Sex and Pornography

David K. B. Nias

Introduction

Sex, lots of it, in the afternoon is the television prediction of psychology professor Kenneth Haun. According to the Daily Mail (13 February 1982), he has predicted that because the number of viewers increases with the amount of sex on television there will be full frontal nudity on day-time soap operas within the next ten years. If this does happen will it be a bad thing? In a review of existing evidence and relevant theories, Eysenck and Nias (1978) concluded that the portrayal of violence is almost always potentially harmful but that the portrayal of sex is not necessarily so. *Fanny Hill* was given as an example of a highly erotic work that could safely be shown on film since any effects it might have on people were unlikely to be of a socially undesirable nature. The tone of the book is one of enjoyment, women are not degraded, and there is no violence to destroy a prevailing sense of good humour and excitement; in other words the theme is prolove, prowomen and prosex. While the portrayal of the many and varied sexual adventures of the heroine might lead to an increase in sexual activity, we argued that it was difficult to see any strong basis for the censorship of such material.

While a case might be made on the grounds of entertainment and education for the portrayal of sex on television, it must be pointed out that many if not most of the commercially available pornographic films are not of a "wholesome", hedonistic character. Rather they are antilove, antiwomen and even antisex (in the sense of a cooperative and rewarding experience). On theoretical grounds this type of pornography might well have adverse effects on viewers, and yet surprisingly the present-day policies on censorship seem to work in favour of such material. Pleasant, enjoyable love-making in a romantic context is rarely shown in films or on television: instead there is an abundance of violence and any sex which is shown is accompanied by an underlying theme of unpleasant- ness or an expectation that the sexual participants should be punished in some way. That such a policy is wrong, emerges not only from a theoretical analysis but also from a consideration of recent research evidence. It is the purpose of this chapter to review this recent evidence on the effects of filmed sex.

The Presidential Commission

Opinions differ widely on the effects of erotica and pornography, as much as in any other area of psychology, and so it is important to keep closely to factual evidence in arriving at any firm conclusions. Only in this way is there any hope of formulating a rational basis for deciding who is right. That personal biases can influence expert judgment is well illustrated by the conclusions of the US Commission on Obscenity and Pornography (1970). This body was appointed by President Lyndon Johnson to survey the existing evidence and to commission new research. The outcome was two conflicting conclusions! The first from a majority of the committee members was that there was no evidence of harmful effects from pornography. This was the "official" conclusion that has subsequently been quoted in courtrooms and in psychology textbooks. The other conclusion, published as a series of minority reports, was that there were numerous grounds for advocating the suppression of sexual materials. These very different conclusions were reached by appointed "experts" who for several years had been studying the same research reports!

The work of the US Commission has been open to much criticism (Cline, 1974). In particular, their theoretical analysis of the problem was very weak. From the field of clinical psychology there are a number of mechanisms, such as imitation, disinhibition and desensitisation, that are used for explaining how people's behaviour can be modified. The Commission made no use of such a background in their analysis of the evidence on how people might be influenced by erotica. Moreover, had the Commission started out with a proper theoretical analysis, then the design of the commissioned research studies would have been more appropriate to the issue in question. They were trying to answer the question of whether or not pornography could have harmful consequences. And yet no explicit attempt was made to design studies that could, on theoretical grounds, be expected to reveal any harmful effects. That such an attempt could have been made has become apparent from research conducted since the Commission.

Imitation

The theory of imitation is well established in the field of learning, and there seems no reason why it should not apply here. The theory postulates that people have a tendency to copy what they see others doing. In technical terms, learning is possible through the symbolic representation of modelled events. Any influence from erotica should thus depend partly on the nature of the material involved. There is a world of difference between loving encounters and kinky violence, and yet the Commission made no clear distinction between material involving normal and abnormal sexual activities. Most of their studies involved normal sex, such as pictures of semi-clad females, and films of love-play and sexual intercourse. Now if the theory of imitation does apply, then we might expect an increase in "normal" sex to follow from exposure to such material. This was exactly what the Commission found and because this increase applied only to normal sex they reached a majority verdict of "no harm".

While exposure to romantic films and to pictures of semi-clad females is unlikely to increase a person's proclivity to rape and sexual perversion, at least according to the theory of imitation, exposure to portrayals of sexual assault and deviant practices might do so. The points is that adverse effects might have become apparent had the research material involved abnormal sexual activities such as rape, bondage and sadism. It has become particularly relevant to investigate this question since, over the past few years, there has been a steady increase in the portrayal of sexual violence in the media. Partial documentation of this trend has been obtained for *Penthouse* and *Playboy* by Malamuth and Spinner (1980), who carried out a content analysis of the pictures in these two magazines over a five-year period. Accompanying this increase in media sexual violence, there has been an increase in sexual attacks on women in the Western world (Court, 1977). Could there be a causal link?

In spite of the apparent relevance of the theory of imitation, and in spite of several studies demonstrating imitation of violence, there have been no studies explicitly designed to demonstrate imitation of sexual behaviour. A possible exception is a study by Wishnoff (1978) who selected 45 sexually inexperienced and anxious women; the State University of New York at Albany. They were randomly assigned to one of three groups. One group was shown a 15-minute film of explicit and varied sexual activity beginning with a couple making initial eye-contact and ending in bed with sexual intercourse. Another group was shown a film of sexually implicit activity with only kissing and fondling being actually portrayed, and the third group was shown a nonsexual control film. Questionnaires revealed a change in attitude on the part of the women in the sexually explicit group. In considering the new sexual behaviours, they were already planning to incorporate them in their future life styles. Wishnoff interpreted this result as due to imitation, pointing out that imitation theory does not require the new responses to be overt at first – it is sufficient to demonstrate an intention to imitate given a future opportunity to do so. The questionnaires also revealed that the women in the sexually explicit film group felt less anxiety after their film than did either of the other two groups. This suggests an alternative interpretation along the lines of the film disinhibiting them. It is difficult on the basis of this study to decide which of these two interpretations is the more likely.

Another study that is open to the same interpretations is by Donnerstein (1980). Male students who had been shown a pornographic film, in which a female was sexually abused and assaulted, were subsequently more likely to administer painful electric shocks to a confederate of the experimenter than were students who had seen a purely sexual film. The students were asked to administer the shocks as part of an experiment on learning and punishment for which they thought they had volunteered; this is known as the "aggression machine" method. The confederate had earlier angered them and they had a chance for revenge when the confederate served as the subject in the bogus learning experiment; the level of shocks to be administered was left to the discretion of the student.

Following the sexually violent film, the shocks chosen were higher if the confederate was female rather than male. Because of this disturbing finding, Donnerstein suggests imitation as a possible mechanism; it appears that the students were associating the female confederate with the victim in the film who was also of course female. This is not a clear case of imitation, however, since the administration of shocks is rather different to the type of assault that was shown in the film. As before, disinhibition theory provides an alternative or additional interpretation.

Disinhibition

The theory of disinhibition refers to conditions under which already existing, but normally inhibited, responses are expressed. Most people are constrained from acting aggressively even when provoked, but given that they have a potential for aggression there may be conditions under which this aggression will be expressed. Several studies have indicated some conditions under which subjects may be disinhibited for both sexual and aggressive behaviour.

Malamuth, Haber and Feshbach (1980b) administered a questionnaire to students at the University of California, Los Angeles, which included items about rape. The majority of the students, whether male or female, believed that most men would rape if they could be assured of not being punished. Moreover, half the male students thought that they too would be likely to commit rape under such conditions. This result is somewhat unexpected since a proclivity to rape is usually considered pathological. Abel, Barlow, Blanchard and Guild (1977) by assessing arousal to audiotaped portrayals of rape and consenting sex found that while convicted rapists tended to be aroused by both types of sex, nonrapists were aroused only by consenting sex (a result also found by Barbaree, Marshall and Lanthier, 1979). It has since been pointed out, however, that pornographic accounts of rape invariably portray the female victim as enjoying the assault, and that this enjoyment may be a crucial variable. In the Malamuth et al. (1980b) study it was noted that the students thought that some women might enjoy the rape experience, although very few of the female students thought that they personally could enjoy it.

Malamuth, Heim and Feshbach (1980c) decided to investigate whether students could be aroused by rape depictions in which the victim experiences an orgasm. After confirming the Abel et al. (1977) finding that nonrapists are less aroused by rape than by consenting sex, they found that this did not apply to rape in which the victim derived enjoyment from being victimised. It was as if her enjoyment disinhibited the students and allowed them to be aroused as much as , or even more than, they would be by consenting sex. For male students, although not for females, this finding applied even if the victim experienced pain along with the orgasm; in fact, most arousal was reported when the story make it clear that the victim experienced both pleasure and pain. For the females, it seems that identification with the victim rather than the assailant might account for their being aroused only when the victim was portrayed as experiencing orgasm without any pain.

Another approach that may illustrate the role of disinhibition involves alcohol, which is well known as a suppressor of inhibitions. Its role in sexual aggression is of particular social concern since rapists often act under its influence, sometimes using it as an excuse for their crime. Rada (1975) found that 57 percent of a group of 77 convicted rapists claimed to have been drinking immediately prior to their crime. In order to demonstrate that alcohol can indeed lower sexual inhibitions, Briddell et al. (1978) presented male students with rape recordings of a female describing consenting sex, forcible rape and sadistic aggression. Self-reported arousal and penile response indicated that most sexual arousal occurred with the account of consenting sex. But after an alcoholic drink, or the belief of having consumed one, arousal levels were significantly higher and there was now as much arousal to the account of rape as to the account of consenting sex. It was as if the influence of the alcohol had disinhibited the students and allowed them to be aroused by material that was normally slightly abhorrent to them. Since this finding applied regardless of the actual content of the drink (an alcoholic and a suitably disguised nonalcoholic version of a malt liquor were used), it appears to reveal a form of psychological rather than pharmacological disinhibition. That confidence can be placed in the result is indicated by a similar effect revealed in an earlier study on arousal to erotic films by Wilson and Lawson (1976).

Evidence that arousal to a rape portrayal could be translated into an aggressive action has been provided by Donnerstein and Berkowitz (1981). Male students were shown a version of a film in which a girl is reading in the company of two men; the men start to make advances to the girl, tying her up and finally raping her. Following exposure to one of the films, the students were given the chance to administer shocks to a confederate of the experimenter who had earlier angered them. Those who had seen a version of the film in which the girl had enjoyed the rape experience tended to administer highel levels of shock. This result applied only if the confederate was female, presumably because of an association with the female character in the film.

The conclusion to emerge from the above studies, and there are others with similar results, is that under certain conditions a wide range of people may be aroused by depictions of undesirable sexual behaviour. Viewed in this light, the fusion of sex and violence in the media is potentially dangerous if it acts to disinhibit aggressively inclined men on the verge of committing rape.

Desensitisation

The theory of desensitisation concerns the capacity of people to become gradually less and less emotional to what initially causes a strong emotional reaction. It is assumed that disinhibiting experiences have a cumulative effect, so that in a sense disinhibition is a special case of desensitisation. Surgeons have to become immune to the sight of blood and to cutting into people, and soldiers are probably more efficient once they get used to the idea of killing. So too it might be argued that people can become desensitised to sexual violence. The US Commission involved several studies relevant to this issue. Sex offenders were compared with control groups in order to see whether they had been exposed to more pornography. It was assumed that if exposure to pornography is harmful then evidence of such exposure should be apparent in sex offenders. In desensitisation terms, prolonged exposure to pornography might lead to deviant acts if the person had previously been constrained by powerful feelings of guilt and anxiety.

The results were conflicting, and are summarised in the report of the US Commission (1970). One study did indeed find evidence of a connection between exposure to pornography and sexual deviance (Davis and Braucht). But another found that sex offenders had less exposure to pornography as adolescents than did ordinary criminals (Cook and Fosen). Similarly another study found that incarcerated sex offenders reported combining masturbation and pornography as adolescents less often than did control groups (Goldstein *et al.*). Finally, yet another study found that sex offenders on probation had a similar level of exposure to pornography as did control groups, but that they reported less exposure prior to the age of 21 years (Johnson *et al.*).

Unfortunately, these studies all suffer from a cause and effect problem that is inherent in the field study or correlational approach. It is possible that sexual offenders deliberately seek out pornography because of their abnormal proclivities. If so then exposure to pornography is more a consequence than a cause of their deviance. Alternatively, it is possible that sex offenders, being "men of action", are less likely as a group to favour looking at and reading about sex. Because of these conflicting possibilities, the above surveys are virtually worthless; results in either direction can be explained away. A clear answer to the question of the effects of prolonged exposure to pornography can be obtained only by conducting an experiment.

The only relevant experiment that appears to have been conducted was by Reifler *et al.* (1971). Volunteers were provided with a wide range of pornographic materials for fifteen days. Relative to a control group, a number of changes in their attitudes were apparent. In particular, and consistent with desensitisation theory, their attitudes to pornography became more permissive as a result of their experience. It is a pity that there has not been more experimentation of this type. If there are substantial numbers of people on the verge of committing sexual offences but who are at present constrained only by their emotions from doing so, and if desensitisation of these emotions can occur as a result of repeated exposure to pornography, then the consequences are quite alarming.

Treatment procedures

Erotica has found a role in sex therapy where attempts are made to "normalise" a patient's sexual behaviour. Sex therapists are typically faced with problems in which patients literally do not know how to go about making love, or in which patients are prevented from so doing by anxiety and guilt. In the case of ignorance, erotic films serve as imitation models; indeed for most people love techniques are based largely on media experience to begin with. In the case of anxiety and guilt, attempts are made to disinhibit and desensitise these emotions. In contrast to much of the research on the effects of erotica, sex therapists have tended to work from established theories in designing treatment. Unfortunately, the success of their procedures has not always been carefully checked. Nevertheless, examples of their work serve to illustrate the possible applications of theory to practice.

Lobitz and LoPiccolo (1972) attempted, with apparent success, to disinhibit inorgasmic women by getting them to role-play "a gross exaggeration of orgasm with violent convulsions and inarticulate screaming". Taboo words have also been used as part of a disinhibiting procedure designed to enrich the sex lives of normal couples (LoPiccolo and Miller, 1975). Herman, Barlow and Agras (1974) attempted to change the orientation of homosexuals by exposing them to erotica involving females and instructing them to develop fantasies of heterosexual behaviour. It may be assumed that desensitisation of anxiety, or the rewarding effects of sexual arousal on new fantasies, might result in therapeutic change. Self-reports and physiological measures indicated that this procedure was sufficient to increase arousal to heterosexual erotica and in some cases to actually increase heterosexual behaviour. Given such a result, it should also be possible for someone to develop a preference for sexually violent pornography.

Individual differences

In the words of D. H. Lawrence: "What is pornography to one man is the laughter of genius to another". In spite of its importance in general psychology, a proper investigation of individual differences in reaction to erotica has only recently begun. One of the best examples is provided by Malamuth (1981), who selected 13 male students who found the idea of rape attractive and compared their reactions to pornography with 16 students who did not.

The students were shown a slide-audio presentation of a scene in which a motorist encounters a female on a deserted road. In one version, they have sex in the car, and in the other version he forcibly rapes her (although she is depicted as finally enjoying it). Following this exposure, the students were instructed to try to reach as high a level of sexual arousal as possible by creating their own fantasies. The most arousing fantasies were generated by students regarded as "force-oriented" from the group exposed to the rape sequence. This result is shown in Fig. 1. Analysis of the fantasies further revealed that five involved sexual violence and, significantly, all five were from students who had been exposed to the rape sequence. This appears to constitute evidence that violent fantasies can be stimulated by exposure to a portrayal of rape, especially in those who are already attracted by the idea of forced sex.

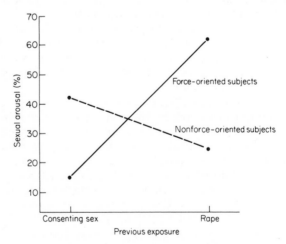

Fig. 1 More sexual arousal accompanied the fantasies of students attracted by the idea of rape after exposure to a portrayal of rape than to consenting sex. (Based on Malamuth, 1981)

Another example of individual differences, this time the perennial gender difference, comes from a study involving exposure to a story in which pain was inflicted on a woman whom, much to her own surprise, experienced intense sexual pleasure (Malamuth *et al.*, 1980b). The students were then asked to evaluate a rape story. Relative to a control condition of exposure to an exciting sexual story, male students who had just read the sado-masochistic passage tended to find the rape story more sexually arousing. But for the female students it was rated as less sexually arousing. In other words, the sado-masochistic story appears to have stimulated or disinhibited the males but to have sensitised or "put off" the females.

Sex-aggression link

Theorists have often posited a close connection between sexual and aggressive feelings, a link that is perhaps best illustrated by the fusion of the two emotions in sado-masochism. Sex hormones are known to be related to aggressiveness, and experiments on electrical and chemical stimulation of the brain show that the areas concerned with the two emotions lie in very close proximity; stimulation of one area frequently results in activation of the other (MacLean, 1965).

The role of erotica in facilitating aggression has been demonstrated in numerous studies (e.g. Baron, 1979; Donnerstein and Hallam, 1978). Conflicting results are apparent in other studies which demonstrate an inhibiting effect of erotica on aggression (e.g. Donnerstein, Donnerstein and Evans, 1975; White, 1979). The discrepant results have been reconciled, however, by showing that different types of erotica are involved in producing the two effects. Facilitation of aggression usually follows exposure to hard-core pornography, aggressive sex and unpleasant themes generally, while inhibition of aggression usually follows exposure to soft-core pornography and generally pleasant themes. This interpretation is consistent with the *Fanny Hill* example quoted earlier and, apart from the above studies, direct experimental evidence has been provided by Zillmann, Bryant, Comisky and Medoff (1981b).

Erotic and nonerotic photographs and film clips were rated in terms of their hedonic valence on a pleasing/displeasing scale. Using the "aggression machine" design, male students were angered, shown a selection of the materials, and then given the chance for revenge against the confederate who had angered them. Most aggression was expressed following exposure to materials classified as both arousing and displeasing. Since an identical effect was obtained with emotionally arousing and displeasing materials devoid of any sexual content, it appears that the effect was due to general arousal and displeasure rather than to sexual arousal and erotica as such.

Having confirmed that facilitation of aggression by pornography may be explained in terms of general arousal and displeasure, Zillmann, Bryant and Carveth (1981a) tried to determine the extent to which aggressive cues were necessary. A sado-masochistic film (regarded as sexual and aggressive) was compared to a bestiality film (regarded as unpleasant but not aggressive). Relative to a control condition and to pleasing erotica, both films were found to increase aggression to a similar degree; the aggressive cues in the sado-masochistic film did not have a detectable effect. The conclusion emerges that it is mainly a feeling of displeasure that results in aggressiveness. In other words, the sex-aggression link is perhaps limited to sexual arousal combined with unpleasantness.

To end this section with the positive effects of "pleasant" erotica, two studies should be mentioned. Jaffe (1981) used the "aggression machine" method and gave students the choice of providing feedback to a confederate by administering shocks when he made an error or by showing him the correct answer. Fewer aggressive choices were made by students who had been aroused by erotica than by students who had seen a neutral film. This indicates that arousal facilitates the dominant response which, for most of us under normal conditions, is of a prosocial rather than an aggressive nature. The other study, by Dermer and Pyszczynski (1978), indicated that after exposure to erotica, male students felt more love for their girl friends than they did after exposure to nonsexual materials.

Ethics of sex research

Lord Longford (1972) led an enquiry into various aspects of pornography in the UK. One section of the report by M. Yaffé dealt with research studies, and this was followed by a statement from D. Holbrook and Mary Miles questioning the ethics of exposing students to pornography. After expressing their doubts about the value of research they wrote: "Nor do we feel that to use students in such tests is justifiable, and we cannot therefore endorse the suggestions for future research". This ethical problem has been recently highlighted in the Malamuth *et al.* (1980c) study, in which students were exposed to a rape story with the female victim experiencing a violent orgasm. If as the authors suggest, "arousal stimuli that fuse sexuality and violence may have antisocial effects", a dilemma is apparent in that they deliberately exposed students to such material. Recognising the problem, they attempted to counteract it by subsequent debriefing. They did this by making a clear statement on the true horror of rape and how it is totally false to imagine that women can enjoy being raped. This was supplemented by a discussion on why this theme is a common feature of pornography, and why such a myth is so prevalent.

Further analyses of this ethical dilemma have been made by Sherif (1980) and by Malamuth, Feshbach and Heim (1980a). Justification for the research seems to rest on pornography perpetuating the myth that women can enjoy sexual violence, on it being socially relevant to investigate any effects that this myth might have, and on there being no scientifically acceptable way of conducting the necessary research short of exposing people to such material. Given this, it has been concluded that debriefing forms an important part of the research design and that it should, in turn, be assessed for effectiveness. The conflict between methodological and ethical considerations is further highlighted here by the observation that to fully investigate the effects of debriefing it would be necessary to withhold it from some of the research subjects! A useful compromise, however, is to compare the effects of debriefing relative to a control group not exposed to sexual materials. Following this procedure there is evidence that debriefing does result in less acceptance of pornographic rape myths (Donnerstein and Berkowitz, 1981; Malamuth and Check, 1980).

Censorship

After concluding that there was little or no evidence of harmful effects from exposure to pornography, the US Commission (1970) recommended to the President: "Legislation should not seek to interfere with the rights of adults who wish to do so to read, obtain, or view explicit sexual material". Faced with the evidence that has since accumulated, they would have to reconsider their verdict. By designing experiments using pornography involving abnormal sexual activities, experimenters have now demonstrated potentially harmful effects. Although advocacy of censorship does not necessarily follow from a demonstration of harmful effects (the cure may be worse than the disease), it is obviously important to ascertain just what the effects of pornography might be.

Apart from the studies described above, there are others that provide circumstantial evidence. Two examples on the effects of fantasy will suffice. An effect of fantasy on behaviour has been cleverly demonstrated by Fenigstein (1979). Male students instructed to fantasise were then asked to select film clips for viewing. Aggressive films were most likely to be chosen following the creation of aggressive fantasies. Carlson and Coleman (1977) asked students to fantasise on sexual themes for ten minutes. Analysis of the content of their fantasies indicated that the most erotic came from those with sexual experience and, at least for females, those with experience of pornography.

Given that people can be influenced by media portrayals of sex, a problem arises mainly with regard to sexual violence and deviance. Contrary to the views of many film and television producers, there is little in the way of evidence that violence is popular. Diener and DeFour (1978) correlated the amount of violence occurring in television programmes with their popularity as measured by the Nielsen viewer index. The correlation was near zero at 0.05. Further, they showed versions of *Police Woman*, either uncut or with violent scenes deleted, to audiences and found that the uncut version was not rated as being liked more. An exception to this finding, however, was that men rated high on aggressiveness did prefer the uncut version. Another exception probably also applies to the portrayal of justified violence, such as when a hero obtains revenge for a wrong-doing. The most challenging exception, however, concerns the portrayal of sexual violence.

It may be noted from Fig. 1, and from the results of Malamuth *et al.* (1980c), that the highest levels of sexual arousal occur in response to violent sex. This combined with the evidence that substantial numbers of male students find the idea of rape attractive (Malamuth *et al.*, 1980b), poses a special problem in view of the increasing tendency for commercial pornography to popularise sexual violence. Does the evidence constitute sufficient grounds for the censorship of such material, and if so could it be successfully enforced? Such questions are outside the scope of experimental psychology, but what can be done is to inform the producers of pornography about the research evidence in the hope that some level of self-censorship will operate. This is probably a forlorn hope, however, if only for financial reasons and an apparent desire to shock on the part of some producers.

When considering sex on television, the research evidence is particularly relevant since considerable censorship already exists. Full appreciation of this evidence will surely lead to a change in existing television policies. Hostile violence and displeasing sexual themes could almost certainly be reduced without any serious loss in entertainment value. While there is little that can be said in favour of any form of violence on television, a case might be made for more in the way of erotic material provided that it is tastefully presented and limited to normal activities.

References

Abel, G. G., Barlow, D. H., Blanchard, E. B. and Guild, D. (1977). The components of rapists' sexual arousal. *Archives of General Psychiatry* **34**, 895–903

Barbaree, H. E., Marshall, W. L. and Lanthier, R. D. (1979). Deviant sexual arousal in rapists. *Behaviour Research and Therapy* **17**, 215–222

Baron, R. A. (1979). Heightened sexual arousal and physical aggression: An extension to females. *Journal of Research in Personality* **13**, 91–102

Briddell, D. *et al.* (1978). Effects of alcohol and cognitive set on sexual arousal to deviant stimuli. *Journal of Abnormal Psychology* **87**, 418–430

Carlson, E. R. and Coleman, C. E. H. (1977). Experiential and motivational determinants of the richness of an induced sexual fantasy. *Journal of Personality* **45**, 528–542

Cline, V. B. (ed.) (1974). *Where do you draw the line? An Exploration into Media Violence, Pornography, and Censorship*. Brigham Young University, Provo, Utah

Court, J. H. (1971). Pornography and sex crimes: A re-evaluation in the light of recent trends around the world. *International Journal of Criminality and Penology* **5**, 129–157

Dermer, M., and Pyszczynski, T. A. (1978). Effects of erotica upon men's loving and liking responses for women they love. *Journal of Personality and Psychology* **36**, 1302–1309

Diener, E. and DeFour, D. (1978). Does television violence enhance program popularity? *Journal of Personality and Social Psychology* **36**, 333–341

Donnerstein, E. (1980). Aggressive erotica and violence against women. *Journal of Personality and Social Psychology* **39**, 269–277

Donnerstein, E. and Berkowitz, L. (1981). Victim reactions in aggressive erotic films as a factor in violence against women. *Journal of Personality and Social Psychology* **41**, 710–724

Donnerstein, E. and Hallam J. (1978). Facilitating effects of erotica on aggression against women. *Journal of Personality and Social Psychology* **36**, 1270–1277

Donnerstein, E., Donnerstein, M. and Evans, R. (1975). Erotic stimuli and aggression: Facilitation or inhibition. *Journal of Personality and Social Psychology* **32**, 237–244

Eysenck, H. J. and Nias, D. K. B. (1978). *Sex, Violence and the Media.* Harper and Row, New York

Fenigstein, A. (1979). Does aggression cause a preference for viewing media violence? *Journal of Personality and Social Psychology* **37**, 2307–2317

Herman, S. H., Barlow, D. H. and Agras, W. S. (1974). An experimental analysis of exposure to "explicit" heterosexual stimuli as an effective variable in changing arousal patterns of homosexuals. *Behaviour Research and Therapy* **12**, 335–345

Jaffe, Y. (1981). Sexual stimulation: Effects on prosocial behavior. *Psychological Reports* **48**, 75–81

Lobitz, W. C. and LoPiccolo, J. (1972). New methods in the behavioral treatment of sexual dysfunction. *Journal of Behaviour Therapy and Experimental Psychiatry* **3**, 265–271

Longford, Lord (ed.) (1972). *Pornography: The Longford Report.* Coronet, London

LoPiccolo, J. and Miller, V. H. (1975). A program for enhancing the sexual relationship of normal couples. *Counselling Psychologist* **5**, 41–45

MacLean, P. D. (1965). New findings relevant to the evolution of psycho-sexual functions of the brain. *In* J. Money (ed.) *Sex Research: New Developments.* Holt, Rinehart and Winston, New York

Malamuth, N. M. (1981). Rape fantasies as a function of exposure to violent sexual stimuli. *Archives of Sexual Behavior* **10**, 33–47

Malamuth, N. M. and Check, J. V. P. (1980). Penile tumescence and perceptual responses to rape as a function of victim's perceived reactions. *Journal of Applied Social Psychology* **10**, 528–547

Malamuth, N. M. and Spinner, B. (1980). A longitudinal content analysis of sexual violence in the best selling erotica magazines. *Journal of Sex Research* **16**, 226–237

Malamuth, N. M., Feshbach, S. and Heim, M. (1980a). Ethical issues and exposure to rape stimuli: A reply to Sherif. *Journal of Personality and Social Psychology* **38**, 413–415

Malamuth, N. M., Haber, S. and Feshbach, S. (1980b). Testing hypotheses regarding rape: Exposure to sexual violence, sex differences, and the "normality" of rapists. *Journal of Research in Personality* **14**, 121–137

Malamuth, N. M., Heim, M. and Feshbach, S. (1980c). Sexual responsiveness of college students to rape depictions: Inhibitory and disinhibitory effects. *Journal of Personality and Social Psychology* **38**, 399–408

Rada, R. T. (1975). Alcohol and rape. *Medical Aspects of Human Sexuality* **9**, 48–65

Reifler, C. B. *et al.* (1971). Pornography: An experimental study of effects. *American Journal of Psychiatry* **128**, 575–582

Sherif, C. W. (1980). Comment on ethical issues. *Journal of Personality and Social Psychology* **38**, 409–412

US Commission on Obscenity and Pornography (1970). "The Report of the Commission on Obscenity and Pornography". Bantam, New York

White, L. A. (1979). Erotica and aggression: The influence of sexual arousal, positive affect and negative affect on aggressive behavior. *Journal of Personality and Social Psychology* **37**, 591–601

Wilson, G. T. and Lawson, D. M. (1976). Expectancies, alcohol, and sexual arousal in male social drinkers. *Journal of Abnormal Psychology* **8**, 587–594

Wishnoff, R. (1978). Modeling effects of explicit and nonexplicit sexual stimuli on the sexual anxiety and behavior of women. *Archives of Sexual Behavior* **7**, 455–461

Zillmann, D., Bryant, J. and Carveth, R. A. (1981a). The effects of erotica featuring sado-masochism and bestiality on motivated intermale aggression. *Personality and Social Psychology Bulletin* **7**, 153–159

Zillmann, D., Bryant, J., Comisky, P. W. and Medoff, N. J. (1981b). Excitation and hedonic valence in the effect of erotica on motivated intermale aggression. *European Journal of Social Psychology* **11**, 233–252

Children's Television and its Responsibilities: The Producer's Point of View

Anna Home[1]

From the beginning I must make it clear that I am writing from the point of view of a producer in the children's department of the BBC. This department is part of the main structure of the television service. Its head reports to the Controllers of BBC1 and BBC2 and ultimately to the managing Director of television. There is no advisory committee or outside watchdog. The Head of Children's Programmes is responsible for editorial policy within his own department and for the scheduling of his own programmes. The Schools Department of the BBC is a separate department, with its own head and production teams. It works within the structure of the BBC, using its facilities and money, but it does not transmit its programmes in the main body of television output.

In addition, it answers to the Schools Broadcasting Council which has final control of content and policy. Schools' television is curriculum-based with overt didactic aims. It is directed at a captive audience and it assumes a back-up of teachers in the classroom and a large body of print material. I do not propose in this article to discuss the BBC's policy of educational broadcasting. I am concerned with television for children in the mainstream of broadcast output.

Children's broadcasting is a BBC tradition. Lord Reith, the first Director General of the BBC, decreed at the very beginning that within a public service system of broadcasting, children should be considered an important audience. Radio *Children's Hour* became one of the most outstanding achievements of broadcasting before, during and immediately after the Second World War. There are many people, including myself, who have vivid memories of series like *Toytown* by H. G. Hulme Beman, Pamela Brown's *The Swish of the Curtain*, the stories of *Norman and Henry Bones*, *Ballet Shoes* by Noël Streatfeild and the classic dramatisation of John Masefield's *Box of Delights*. In addition to these dramas, there were

---

[1] The author is no longer working for the BBC.

factual programmes about nature, history, etc. Some people would argue that these classic productions were far superior to those we provide for children today. Whether that is true or not, it is certain that radio *Children's Hour* had considerable influence on the generation of children brought up on it. Whether or not we learned anything from it is another matter. It depends of course on the definition of "learning". I certainly learned that I enjoyed stories and drama, and I found out what it was like to go to a ballet school. I also realised that music and sound effects could be used to heighten tension and that fear could be exciting. I suppose I learned that on the whole, good was better than bad, but I can't remember a good deal about prosocial values or teaching points. However, I am absolutely certain that the morality of those *Children's Hour* programmes was crystal clear.

When television became a reality in the 1950s, children's television was automatically built into the schedules. It was now a visual medium, so there were puppets as well as people and *Muffin the Mule* was created, but there were also information programmes like *Blue Peter* and drama, mainly the classics, e.g. *The Secret Garden*, the stories of E. Nesbit, historical fantasies and much comedy. It was all very much in the tradition of radio *Children's Hour* – safe, solid and unworrying – Auntie BBC would maintain the standards and the traditions. However, the advent of commercial television caused a reappraisal of traditional attitudes. Suddenly television was a great deal brasher and less middle-class. ITV companies originated their own children's programmes, as they still do, but they also bought-in American cartoons and started to woo the child audience with programmes of a somewhat more popular nature than the staid BBC. Children had become part of the commercial world.

It was in the 1960s with the enormous expansion of television transmission time, its popularisation and the huge increase in viewers, that people started to worry about the effects of television, particularly on the young. There was a school of thought which was convinced that we were breeding a generation of square-eyed illiterates who would never open a book. Instead they would roam the streets armed with offensive weapons and carry out muggings as seen on the small-screen. The early research was imprecise and inconclusive, and sadly this is true of much which has followed. However, despite the criticism, the BBC continued to expand and develop its children's output. Specialised programmes for the under-fives like *Play School* were created, and story-telling programmes like *Jackanory*. Both of these have now been running for nearly twenty years and have become accepted classics of the children's television world.

There has also been a steady move towards innovation, pushing out frontiers, not in an aggressive or revolutionary way, but slowly and steadily. In the ten years that I have been associated with children's drama, we have come a very long way. We are able to tackle subjects now which we would not have considered ten years ago. For instance, in the last series of *Grange Hill*, there was a perfectly serious discussion of the problems of girls starting menstruation. This is a subject which no one would have dreamt of tackling in the past. Attitudes to children's television have changed as well. In the early days, although everybody agreed that programmes for children should be done, there was a tendency to think that they could be done more cheaply and with less time and resources than those made for adults. Although there is a slight tendency for those who control the purse strings to continue to think in this way, it is much less than it was. People now realise the enormous significance of television as far as the young are concerned.

In the early 1980s the whole of television is in a strange position – it stands at a crossroads. Inflation has hit the BBC very hard indeed and is starting to have an effect on the independent television companies, yet we are about to have a second commercial channel. Equally we are on a brink of a new era of communication with home video, satellite and cable television. In twenty years time the traditional television set up as we know it today will be a thing of the past. All this is going to have an effect on how children view, what they see and what they learn from it. However, for the moment let us take television as it is today and consider it in relation to the children's audience. It is undeniable that television is a major influence in a child's life. In the UK children watch an average of 20 hours a week. However, it is very important to remember that television is only one of a number of major influences on a child, others being school and parents. People often seem to overlook the fact that television is only part of a general social scenario and cannot be considered in isolation. The other thing that has to be remembered is that for today's child television is something perfectly normal, it is part of the furniture; it is not something special, new or different, which it was to the middle-aged generation who tend to fulminate about its effect on children.

Of the hours the average child watches, a considerable number are likely to be of adult television rather than programmes specifically aimed at a child audience. Here lies one of the great problems for television planners. In the UK there is a 9 p.m. watershed, i.e. all programmes transmitted before this time in the evening are supposed to be acceptable to a child audience, but this watershed is often a myth, particularly at weekends. It can be seen from all recent research figures that large numbers of children are watching between 9 and 11 at night and on Saturdays and Sundays even later. But should the fact that those children are still there, prevent the scheduling of controversial adult material, or should the responsibility be the viewers'? Is it really fair that the innovative drama producer should have to worry about upsetting nine-year-olds? Equally, if children are watching so much adult television, is there any point in having a special children's department? In America, apart from a few outstanding and honourable exceptions, the transmissions for children are hardly distinguishable from the so-called family viewing. Surely specialist children's programmes are such a drop in the ocean of the overall output, that we might as well not bother?

I believe that the existence of a separate children's output is vitally important. Children do have specific needs; they are learning and developing and they do have a literature and a culture of their own which should be reflected in their own special television. A five-year-old is different from a nine-year-old and a nine-year-old from a thirteen-year-old. There is not much difference between a thirty-year-old and a forty-five-year-old in programme terms. I believe that it is essential to have a department made up of expert programme makers who are informed about and concerned about their particular audience. They do not have to be specialists in child development or child psychology, but they need to take time to study their audience and its particular needs and to do their homework in terms of the research available. The main aim of a children's department is to produce programmes directed at a very specific audience. These programmes are primarily intended for entertainment – after all, the main output for children is in the afternoon after school – this is a time when they want to relax as much as we do when we come home from work. However, there is no reason why entertainment should be mindless, and the children's producer is generally trying to inform and stimulate within the entertainment format. Children's television is now one of the bigger BBC programme outputs; 850 hours a year, and still expanding. It is a microcosm of the adult output with its own news, documentaries, drama and comedy, and as I mentioned earlier, special programmes for pre-school children. There is no question of this output being intended as a baby minder or as moving wallpaper; it has positive aims, but how positive should we be and what is our function? This brings me back to the theme of this book – television and learning – I think that everybody working in children's television believes that children can and do learn from television. Therefore we should provide them with something worth having. Equally we do not believe that it is our function to teach in the traditional sense. That we leave to our educational colleagues. There is, however, a very fine line between stimulating ideas and discussion and handing out dollops of received dogma. These dangers are well illustrated in some American teenage drama series. The intentions are laudable, to provide realistic contemporary drama relevant to today's audience.

Sadly, the end product is a heavy-handed documentary with the problems and the prosocial values overloading the story line. The result is that the entertainment value falls, the audience feels it has been "got at" and nobody watches. We in the BBC have tried to avoid this in the drama series *Grange Hill*, which has now been running for five years. This is a contemporary series set in a realistic and familiar background, i.e. a school. It is deliberately constructed in a "soap opera" style and it goes out twice weekly in that recognisable format. Its primary aim is entertainment and its secondary aim is to raise issues and provoke discussion. We have dealt with some very serious subjects in *Grange Hill*, ranging from child molestation through race relations to death. We have tried to deal with these subjects so that they arise naturally from the story line and treat them in such a way that they are not rammed down the viewers' throats. I am aware that many people consider *Grange Hill* to be a negative and dangerous programme, and that if children learn anything from it, it is violence and bad manners. Obviously, I do not believe that; rather, I hope that maybe some of the audience learn that other people have the same problems they do; that bullying doesn't pay; that lessons can be enjoyed and hopefully some of them find, when they happen to watch with their parents, that subjects are raised which they can then discuss together, which they have never been able to discuss easily before.

Grange Hill is a positive programme, and its morality is as clear as in those traditional *Children's Hour* radio programmes – good is better than bad, right triumphs over wrong. However, in order to demonstrate that morality, the bad has to be shown in action, and that is when the complaints begin.

I do not believe, and my conversations with children bear this out, that the average child believes that what goes on in *Grange Hill* could go on exactly like that in his or her school. He knows it is fiction, that events have to be structured in order to create a dramatic plot. Perhaps we have something to learn from children's attitudes to television. I think that they may be far more selective and critical than their adult counterparts.

The success of *Grange Hill* has led the BBC to reappraise its attitude towards "teenagers". This was an audience which, until recently, had been largely ignored. The theory being that people stopped watching specialised children's programmes at about the age of twelve and there after began watching adult programmes. However, it has become patently apparent that there is a lost audience which does require programming specifically made for it. The thirteen- to twenty-year-olds are an increasingly important audience with an increasing number of problems and as the problem of youth unemployment continues to spread, it will become even more so. The BBC has initiated a number of programmes in the early evening on BBC2 to cater for this audience. *Maggie*, based on the novels of Joan Lingard being the first drama in this slot, and it is intended to expand this type of programming. ITV will do the same, although its possibilities are limited by a much more compli-cated scheduling problem. The intention of this type of programming is to reflect the world of teenagers, not necessarily to answer questions or solve problems, but to lay out the various issues and discuss them in a way which may be positive and helpful. This audience is one which is particularly sensitive to being "got at", and it is difficult to help them. However, I am absolutely certain that this is a most important area and one which needs much attention paid to it in the future.

What are children learning from watching children's television? Under-fives learn nursery rhymes and songs and finger play. Older children learn how to make things out of egg boxes, etc., to draw things, to write things. They learn about others less fortunate than themselves. *Blue Peter* every year has an appeal for some good cause and through it children will learn about the plight of the starving in Ethiopia, or the disabled in their own country. Some viewers learn, as I did from *Children's Hour*, that stories that are told or acted may also be read – they exist in books. I am glad to say that the idea of the square-eyed nonreading child is dying. It is now acknowledged that a book which is serialised or read on television will be borrowed from the library, or bought. There is now great cooperation between the world of children's television and children's books. Novels are reissued with television-tie-in covers, and at children's book fairs, television story tellers are a great attraction. *Jackanory* is the classic television story-telling programme. It goes out for fifteen minutes a day, five days a week, for more than half the year, and it concentrates on the straightforward narrative. All depends on the story teller. The pictures which are there are in fact incidental to the text. It is interesting that *Jackanory*, once decried as televised radio and a minority programme for middle-class bookworms, is in its sixteenth year and going strong. It is also worth noting that this programme never set out to teach anything about reading or books, merely to tell the best stories in the best possible way. No one *need* learn anything from it, other than how enjoyable it is to be read to, but many children do learn the positive pleasure that comes from books.

News was once something which was not presented for children, except in the chatty form of children's newsreel with its reports of prize horses and maypole dancing. Somehow it was felt that children should be protected from the reality of life (when I was at school we weren't allowed to read newspapers until we were well into our teens), but the news, the adult news, is one programme seen in most households, often immediately after the end of children's television transmission. Frequently the main stories seem to spring from nowhere and if that is true for adult viewers, how much more so for children? It was therefore decided to create a news bulletin for children centred around a presenter who was already familiar to them, John Craven. *John Craven's Newsround* explores the background to stories and puts them in their context. Although it features items of particular interest to children, it also covers main headline stories, often scooping the early evening news. It also follows stories up; so often when a story drops out of the headlines it disappears for ever.

Then there are documentaries and music programmes and science programmes like *Think of a Number*, and sports programmes and multi-faceted magazine programmes like *Blue Peter* and *Multi-coloured Swap Shop*. (I am using BBC examples, but all these programmes have ITV equivalents.) Children will learn something from all these programmes. In fact, probably the only programmes from which the audience learns almost nothing are American cartoons, which are, of course, among the most popular programmes transmitted in children's viewing time. Many people consider these programmes totally mindless, and therefore believe that they should not be transmitted. I think they are no more mindless than many an adult light entertainment show and equally enjoyed. I think that we, as adult providers of children's television, must be careful not to take too moralistic and puritanical a stand. A little rubbish is an excellent thing (in time, the kids may even come to realise it is rubbish and that is learning). The essence of BBC children's programming is balance and variety: it endeavours to cater for every age of child and every taste, and it tries to provide a coherent context for life. But as I said earlier, television is only one aspect of every child's life, and television alone cannot provide everything that a growing child needs for its development.

I would now like to discuss the producer's responsibility and will use the drama output to illustrate my argument. All television producers have a responsibility towards their audience. It is the audience which provides them with their *raison d'être*. However, the producer of children's television programmes has a far greater responsibility because his audience is immature, impressionable and vulnerable, and what is more, his audience has no power and no political impact. In theory, it would be possible for an irresponsible producer to disseminate a steady stream of propaganda using a dramatic vehicle. In practice, the system of departmental responsibility would make it impossible. Nevertheless, producers never forget the enormous power that they hold.

As I have stated earlier, I believe my primary aim is to entertain and my secondary one to educate and inform. How do I set about this task in terms of drama programmes for children? My choice of material is determined by the number of slots I have to fill and by the money I have to spend; drama is the most expensive form of television programming. I also have to consider the needs of the audience and decide how best I can cater for a wide range in both terms of age and interest. I look for variety in both content and style and I look for quality. I also look for something that will challenge both the programme makers and the audience. I try to reflect classic children's literature and the best in what is new. I also try to find new young exciting writers and commission original material for television. At no point do I consider whether my audience is likely to learn anything in the strict sense. I did not choose to dramatise Katherine Cookson's *Our John Willie* because the audience could learn about the conditions in the coal mines in the 1850s, but of course, they did learn something of that. Nor did I choose Bernard Ashly's *Break in the Sun* because it would teach children about the problems of step-parents and incontinence, although many children will have seen parallels with their own lives and gained some comfort. I chose both stories because they were good, quality stories with the power to stretch the minds of the audience as well as to entertain.

Once having chosen the book or story line, what is the producer's next responsibility? Again, to remember the needs of the audience and to make judgments on how far to go. For instance, in *Our John Willie* there is a pit disaster, a mine is flooded; it is a very exciting, dramatic incident, and we decided to treat it as realistically as possible. We also decided that we had to show the aftermath of such a disaster. Therefore, we had to decide how much of the true horror we could show, remembering that at 5.15 p.m. on Wednesday afternoon, there would be quite young children watching. The sequence we eventually shot was realistic and horrific and sad, but not in a macabre or distorted way. It showed that men died and women wept, but it didn't linger on horrific detail. This is the kind of judgment which children's programme producers must exercise. I believe children watching that programme would have been involved and moved – I do not believe that they would have had nightmares afterwards.

However, this belief is not based on concrete research; between 6 and 7 million children watched *Our John Willie* and I know that many of them enjoyed it because they wrote and said so, or commented to people who passed it on to me. Librarians tell me that the book is in constant demand. Other than that, all I can offer is a kind of "buzz", a feeling one gets when a programme is really taking off; it becomes part of the current school and playground gossip. This is not very scientific, I grant you, but I think all of us who work in this field would be glad of more research to back up our hunches. It would be interesting to know more about what programmes children want and why.

Recently there has been a certain amount of research done into *Grange Hill* and it has become patently apparent that this is currently the most popular children's programme in the UK among all age ranges from five to fifteen. It is more difficult to find out why, but the main impression seems to be because it relates to a world which children understand and which is very immediate to them. Research with children is extremely difficult and it is notoriously hard to persuade them, particularly when they are very young, to give honest answers to the questions that are posed; they tend to tell you what they think you want to hear. However, much of the research into children's television seems to have been directed at the negative rather than the positive aspects and has been of a very theoretical nature far removed from the realities of programme making. Hence, the somewhat strange relationship between researchers and practitioners. In addition, we have the awful warning of America before us, where there are innumerable research projects and advisory committees, and fewer and fewer good programmes. Programmes are not made by committees or advisers or researchers. They have to be made by committed programme makers who make their own decisions and judgments. Of course, there is a place for research, but as an aid to the programme maker, not as a negative restraint.

Children and young people can and do learn from television and those of us who make programmes specifically for them must continue to be aware of this. Our task is the most difficult and perhaps the most important in all television. We have an audience which is both vulnerable and impressionable and television is a huge influence in their lives. The 1980s are going to see an increasing need for programmes for the fifteen- to twenty-fives, for school leavers who are not going to find jobs. The rioting that has recently happened in Brixton and Liverpool and the attitude of the young to the police, all these are symptoms of a restless and unhappy decade and television must try to help. The young are going to need programmes that will be relevant to them both in terms of continuing their education, but also which will help them understand their predicament and enable them to relax a little. In the 1930s the cinema boomed, perhaps the same will happen to television in the 1980s. We can do much through the medium of television, but unfortunately we cannot provide the complete answer.

If I were asked to give advice to a young producer starting to work in the field of children's television in 1983, I would say "First of all, think of your audience and its specific needs, then enjoy what you do and do it wholeheartedly, with honesty and humour".

Learning How to be Intelligent Consumers of Television

Dorothy G. Singer and Jerome L. Singer

Problems and possibilities of the television in our homes

Television is a member of the family. Ensconced in 99% of American homes, the television set claims the attention of school children for approximately $4\frac{1}{2}$ hours per day reaching a peak of $5\frac{1}{2}$–6 hours a day by age 12 (Lyle and Hoffman, 1972). Poor children watch more television than middle-income children (Greenberg, 1976) and black children watch more than white children (Greenberg and Dominick, 1970). In one of our major studies at the Yale Family Television Research and Consultation Center we have found that even middle-class and lower-class nursery school children averaged about 22–23 hours per week viewing television, and for our middle-class sample, the range of viewing per week was from 1 hour to 72 hours (J. Singer and D. Singer, 1981). Obviously, television is playing an important part in children's lives, and in America, a large segment of the school-aged population is spending more time in front of the screen than in the classroom.

Social scientists and educators view television as a villain, and indeed upon examination of the current research related to reading habits (Zuckerman, Singer and Singer, 1980b), imagination (Singer, 1982) and enthusiasm in school (Postman, 1980), this charge is not unwarranted. The fast pacing of American television, the quick resolution of problems, the interruptions of a story by commercials, the use of special camera effects that blur reality and fantasy distinctions, all pose potential problems for the child in the classroom. It has been suggested that young children have a distorted comprehension of television programs and are often confused about the sequence of events in a plot (Collins, 1975; Collins, 1978; Meyer, 1976). Older children are more sophisticated about the medium and learn its conventions, but even at the third grade children "pay attention to and remember fewer or different aspects of programs than adults would expect" (Collins, 1978, p. 199). With experience, children are able to develop a better understanding of television content and eventually learn to evaluate and criticise what they view. If adults, parents or teachers, however, interact with young viewers and assist them in interpreting events, in clarifying the distinction between reality and fantasy and in analysing programs, children can more readily understand a program and use the content in a constructive way (Singer, Singer and Dodsworth-Rugani, 1979; Tower, Singer, Singer and Biggs, 1979; and J. Singer and D. Singer, 1981). Teachers are especially important as mediators when television is used in the classroom for instructional purposes.

Some attempts at instructional uses of television

Two of the leading guides for teachers and parents are *Prime Time* and *Teachers' Guide to Television*. Both are published privately, but receive network support. Each is geared to the network specials and presents ideas that teachers or parents can use to help junior high and high school children increase and expand their knowledge of the subject matter shown. There were guides prepared for *Roots*, *The Scarlet Letter*, and such afternoon specials as *Snowbound*, *Luke was There*, and *Papa and Me*. *Prime Time* is generally circulated through the schools, and parent groups and school groups can subscribe to *Teachers' Guide to Television*. Unfortunately, aids such as these have not yet been evaluated, therefore we have no empirical data on the usefulness of these projects.

The Agency for Instructional Television, a consortium of nineteen state and provincial agencies, assists education through the development of cooperative television program projects for use in the schools. The programs concentrate on materials designed to promote essential skills such as critical reasoning and study skills, as well as television programs for science, mathematics and reading. These lessons are used by some school systems via cable.

Publishers are also becoming more interested in developing workbooks for use in language arts curricula that focus on television stories and characters in order to motivate students to develop critical thinking skills, increase vocabulary, learn correct grammatical usage, summarise, use analogies, and in general develop expressive language (Singer and Stoving, in press; Potter, 1976; Potter, Faith and Ganek, 1979). For preschool settings, there are guides developed by Appalachia Educational Laboratory in West Virginia for use with *Mr. Rogers' Neighborhood* or *Captain Kangaroo*.

QUBE, a cable experiment piloted in Ohio, which permits families to respond to the program host by pushing buttons on their set is still being evaluated. The belief is that through this two-way system people could be polled immediately about an issue. When the system is used as part of an instructional program, students and teachers can interact as if they are face-to-face in a regular classroom.

There are, of course, columns in newspapers that deal with television, and some actually give advice to parents about how television can be used constructively (Potter, 1981; D. Singer and J. Singer, 1981–1982). Magazines for children such as *Cricket* and *Scholastic* also have features relating to television. In addition to these, several church groups issue newsletters for parents or Sunday Schools suggesting positive uses of television.

Finally, the networks themselves are doing some things to help parents and children. There are public service announcements that suggest that parents and children view programs together and discuss them. They suggest too that parents *select* programs for their children and use a guide or newspaper for descriptions of appropriate programs for their family. The networks, specifically CBS, have supplied television scripts for a reading project in Pennsylvania, and ABC has made scripts of *Roots* available for classroom use. NBC has spots that deal with how special effects are produced. Programs such as *Footsteps* aired on Public Television that deal with parenting, or documentaries or specials that touch on educational subjects, demonstrate ways in which television can not only be entertaining, but useful in providing emotional and social content. Another program on Public Television, *Freestyle* was designed to counteract social stereotypes, and is aimed at the preadolescent viewer. The various news format programs on Public Television prepared for children are designed to bring information about world events to this younger audience in a way that is interesting and relevant to their lives. One such program, *Why in the World*, began in the Fall of 1981 and is directed towards a high school audience. (Study guides to accompany each segment have been made available for classroom use.)

Persisting issues: studies of stereotyping and aggression

Despite these attempts to use television as an educational medium, some researchers are still concerned about the inability of very young children to separate fantasy and reality on television. Schramm, Lyle and Parker (1961) suggested that fantasy is the main reason why children watch television, but work by Dembo (1973), Dembo and McCron (1976), Greenberg (1974), von Feilitzen (1976) indicates that children use television for more than escapism and as a source for fantasy satisfaction. Children not only learn ways of behaving, dressing, but they get general information about different parts of the world from television. What children actually understand and how they utilise the information is dependent of course on their age and their intelligence. It is increasingly apparent in the United States that television is becoming credited as the "most believable news medium", even more than newspapers, radio or magazines (Roper, 1980). Bearing this in mind, not only is it imperative to teach children how to understand how news is prepared and edited for television, and how to be able to interpret and analyse television news as compared to radio or print, but it is important to teach children how *other* social messages are conveyed in nonnews programming (Singer, Singer and Zuckerman, 1981).

Given that television affects our imagination (J. Singer and D. Singer, 1981), information and reading (Hornik, 1978), our view of women (Lemar, 1977; Cantor, 1978; O'Bryant and Corder-Bolz, 1978), our attitudes towards minorities (Greenberg, 1982) and even our purchasing habits (Liebert and Schwartzberg, 1977), a course in understanding television would seem efficacious in remedying false assumptions concerning the mechanics of the television medium (simple electronics, special effects, camera techniques), program content of television, and television commercials.

The amount of television viewing, particularly heavier viewing of action-adventure programs, has been found to relate significantly to children's aggressive acts in nursery school (J. Singer and D. Singer, 1981). Viewing television violence has also been demonstrated to be a significant influence on the occurrence of aggressive behavior in elementary school-aged and adolescent children (Busby, 1975; Eron, 1980; Lefkowitz, Eron, Walder and Huesman, 1977; Surgeon General's Scientific Advisory Committee on Television and Social Behavior, 1972). It is conceivable that if children understood that television violence is "pretend" using highly trained stunt people to perform dangerous acts, or actors trained to pull their punches, and that props are designed to break easily, that music and lighting create suspense and excitement, there would be less inclination to imitate the aggressive acts on television.

A study of family viewing, prejudice and imagination

In a study carried out with more than 200 third, fourth and fifth grade children (ages nine to eleven), we were interested in whether their weekly television viewing would be associated with race prejudice and sex prejudice (Zuckerman *et al.*, 1980a). Research indicates that children who watch television more frequently have more traditional views of sex-role behavior (Beuf, 1974; Frueh and McGhee, 1975). On commercial prime time television and even on public television, men outnumber women in major portrayals, and are engaged in a wider variety of occupations (Greenberg, 1982). Blacks, Hispanics, Asian-Americans, and Native-white Americans appear in programs less frequently than Caucasians and are usually presented in lower-level occupations (Greenberg, 1982).

Using parents' questionnaires, parents' records of family television viewing and the Katz–Zalk Prejudice Test of racial and sex prejudices, we assessed the amount of prejudice in our sample and its relation to television viewing habits and demographic variables (including IQ and reading scores); data were available for 155 children. Children were relatively light viewers (averaging 15.1 hours per week) when compared with the national average of 20 to 30 hours per week of viewing time for children in this age group (Comstock, Chaffee, Katzman, McCombs and Roberts, 1978). The types of programs they watched, however, were typical of the national group. Their records indicate that they viewed on average 3.6 hours of comedies, 3.0 hours of sports, 2.8 hours of cartoons, 1.8 hours of action or violent programs, 1.5 hours of game or variety programs, and 1.2 hours of dramas. The sports average is higher than usual because records were kept during the World Series period. Of these programs, the sample averaged one hour of fantasy or violent (non-cartoon) programs, 0.4 hours of programs depicting extreme stereotypes and 0.2 hours of programs with primarily black characters each week. Educational programs were watched by very few children. The parents were also light viewers; mothers averaging 10.4 hours per week and fathers watching 11.0 hours each week. Of this time, 1.5 to 2.0 hours were spent watching action or violent programs.

Results indicate that girls who watched more game/variety programs and reruns that are demeaning to women, such as *I Love Lucy* and *The Jetsons*, were more prejudiced against girls than their classmates were. In contrast, the girls who watched more fantasy or violent programs and *fewer* other violent programs were less prejudiced against girls. The role models in fantasy or violent programs are competent and beautiful women such as *Wonder Woman*, and are *not* the victims who are usually found in action-detective shows. Children who were the most prejudiced against black children were those who watched more violent programs (where blacks are portrayed in negative ways) and *fewer* programs with major black characters such as *What's Happening* and *Diff'rent Strokes*, where blacks are portrayed more favorably.

These findings do not *prove* that television causes prejudice, but the results do suggest that some kinds of television programs may tend to encourage or discourage prejudicial attitudes.

Reading and television viewing

We were also interested in the relationship between reading and the television viewing patterns of our sample (Zuckerman *et al.*, 1980b). One explanation to account for a decrease in reading by heavy television viewers is the "displacement" theory which suggests that television viewing "displaces" reading. Watching television has become such a popular pastime that many children have little time for reading, and less incentive to learn to read (Hornik, 1978). During the elementary school years, when children are first learning to read and need to practice reading in order to improve their skills, they tend to spend more and more time watching television. Displacement is related to "gratification" theories, which assume that television replaces reading because both activities satisfy similar needs (Blumler and Katz, 1974). According to these theories, during the pretelevision era, children and adults sought diversion or excitement through reading "light fiction" or "pulp" magazines, thereby practicing their reading skills; even comic books provide some reading practice. In contrast, television provides entertainment with a minimum of effort on the child's part. The hours that used to be spent reading are now taken up by television viewing. Obviously, children who spend at least 35 hours each week at school and 20–35 hours each week watching television have little time to spend reading at home.

From a physiological viewpoint, television viewing may have a deleterious impact on reading because the two activities involve different brain functions. In most right-handed people and some left-handed people, visually presented material is apparently processed primarily by the right side of the brain in a global fashion. Printed material or complex verbal or mathematical sequences seem to engage primarily the left side of the brain. Although reading involves both visual and cognitive processes, some researchers have expressed concern that television viewing is enhancing a strong preference for or reliance upon global visual representations. As a result, children and, later, adults will probably be less patient at making the effort required to process auditory–verbal material, such as teachers' lectures, or to deal with reading material. Males appear to be more differentiated in right- and left-brain functioning than are girls (Witelson, 1976) which may account for the fact that boys are much more likely to have reading problems than are girls. Research indicates that boys watch more television than girls (Singer, 1979), and are therefore exposed to more visually oriented material in which the verbal component is presented very rapidly. It is possible that this visual emphasis tends to increase boys' difficulties in developing verbal skills. In addition to studies on hemispheric differences, research on the amount of brain activity suggests that there is more extensive and diffuse brain activity during reading then during television viewing.

Although television may satisfy some of the same needs as reading, and may also stimulate cognitive or imaginative skills, in general, the task of reading involves a more complex set of processes than the encoding of the visual images on the television screen. When reading, the child is confronted with a much more difficult set of transformations. Printed symbols must be transformed into sounds and grouped and organised into meaningful words, and these in turn must be organised into sentence groups. The next step in the process inevitably involves some degree of imagery representation. The child who reads a phrase about a boy who finds a new stuffed animal on his bed when he comes in from play must try mentally to reconstruct that situation in some form. Thus, he or she creates a series of images, and even if the book itself has some pictures to help, the reader must elaborate on the script by filling in the details.

As children grow up and read more extensively, pictures play a much less prominent role in the written material, and the child must use his or her private sets of images to fill in. Therefore, reading necessitates an active stance by the child; indeed, there is research demonstrating that the efficient reader is an active thinker when reading (Blumenthal, 1977). The reader builds up configurations in the short-term memory system and then scans the text for inferences that support the developing configuration.

Since reading is a basic skill for effective functioning in society, we have to be concerned about the possibility that heavy television viewing may interfere with the child's development of reading skills. Research conducted in the 1950s and the early 1960s tended to find no significant relationship between television viewing and grades (Greenstein, 1954; Ridder, 1963). More recent studies of the relationship between reading and television viewing sometimes have found significant results, but these studies did not control for IQ or socio-economic status (Witty, 1967). When IQ or socio-economic status were controlled, the relationships between television viewing and school achievement were no longer significant (Furu, 1971; Thompson, 1964).

One recent study provides evidence that television may interfere with reading; this study found a negative relationship between television exposure and long-term reading skills growth (Hornik, 1978). However, these data are based on a comparison between children in El Salvador whose families did or did not own television sets. Although social class was discussed as a potentially confounding variable, it was not statistically controlled in the analyses. Moreover, the impact of the introduction of television into a household appears to be different from the impact of television on children who have lived with it for several years. Therefore, the results may not correspond to those for children in the United States.

In addition to the obvious relevance of parents' educational attainment or related socio-economic measures, there are parental behaviors that may influence children's television or reading habits, such as the parents' own television viewing and reading habits; yet these variables have not been assessed in the research on children.

In addition to the information we collected about the programs our sample watched, teachers rated the children on imagination, attentiveness in class, and enthusiasm in class. The reading scores were obtained from a standardised test and were at grade five plus six months, for this sample of third-, fourth- and fifth-graders; the mean IQ was 110. In one school, the teachers kept records of the number of books that children read during a four-week period. The number of books read by 94 students during that time averaged 4.8, and ranged from 0 to 24 books. These figures were not considered to be precise measures of reading, since children who read more books may be reading shorter or less sophisticated books. This confounding factor was controlled to some extent by including grade level in a regression analysis. Children in lower grades read more books; when grade was controlled, viewing fewer game or variety programs predicted reading more books.

The amount of time that children spent reading on an average week day was predicted by three of the variables assessed. Children who spent more time reading had higher IQs, had more highly educated fathers, and watched fewer fantasy or violent programs. Imaginative behavior was produced by higher IQ and viewing fewer fantasy or violent television programs. Attentiveness in class was predicted by higher IQ and female sex but was unrelated to television viewing. Enthusiasm in class was predicted by higher IQ, female sex, child's viewing fewer cartoons, and child's heavier weekly television viewing.

Mother's educational attainment, mother's employment status, children's birth order and number of siblings, parents' television viewing, the number of television sets in the home, and parental limits on television viewing times were not significantly related to IQ, reading ability, reading behavior, or school behavior. Children's total viewing time and their viewing of comedies, nonviolent dramas, cartoons, or public television programs were unrelated to IQ, reading ability, or reading behavior.

The results suggest that fantasy or violent programs may inhibit or take the place of imaginative play or imaginative behavior. This is consistent with research on preschool children (J. Singer and D. Singer, 1981). Since imagination was related to IQ and cognitive growth, this is a potentially important finding that warrants further research. Enthusiasm in school was predicted by watching fewer cartoons and by heavier overall television viewing. Of all the programs on television, cartoons are probably the most visually oriented. Since IQ was controlled and age was not a significant factor, the inverse relationship between viewing cartoons and enthusiasm may reflect impatience with a relatively calm and bland school environment. However, it is important to note that overall television viewing of about two hours per day did *not* have a negative impact on any of the behaviors that we assessed and was, in fact, positively related to enthusiasm in school. It is especially interesting that attentiveness in class was unrelated to any particular kind of television viewing tested.

The data do *not* prove that television viewing influences reading habits, imagination, or enthusiasm for learning. However, the results clearly suggest that particular kinds of programs may interfere with or take the place of reading and imaginative behavior. In addition, the results indicate that the children who watch more cartoons are rated by their teachers as unenthusiastic about learning. If cartoons are not causing decreased enthusiasm for learning, then cartoons must have a special attraction for children who, regardless of IQ or socio-economic status, are "turned off" by school. In either case, this relationship should be examined further.

Why did there appear to be no relationship between reading ability and television viewing habits? One explanation is that the children in the sample were relatively moderate television viewers, averaging approximately two hours per day. This leaves a considerable amount of time for reading. Moreover, (the schools that the children attended heavily emphasised reading skills, so that the children spent a great deal of their school time participating in reading activities.) The extensive school reading program may also account for the lack of a significant relationship between parents' educational levels and children's reading scores, and the very modest relationships between children's IQ and reading scores or parents' educational attainment. If a school is successfully teaching reading, regardless of children's IQ or socio-economic status, it is not surprising that children's home behaviors do not significantly influence their reading ability.

In contrast, although the children's reading habits were not significantly related to their total weekly television viewing, they were related to the *types* of programs they watched. The amount of time the children spent reading was inversely related to the amount of time they spent watching fantasy or violent programs, and the number of books they read was inversely related to the amount of time they spent watching game or variety programs. The explanation that seems most appropriate for these findings is that of gratification. It may be that fantasy or violent programs provide the same kinds of excitement as fairy tales, adventure books, comics, and other popular children's books, and therefore satisfy similar needs for escapism and fantasy. The results are therefore consistent with Murray and Kippax's recent data which indicated that watching television tends to replace comic books (Murray and Kippax, 1978).

Perhaps one of the most interesting findings in the study concerned the relationship between parents' and children's television viewing habits. In evaluating this relationship using records kept by parents, and demographic background (age, grade, sex, birth order, number of siblings, and parents' educational levels) we found that parents' television viewing habits were the most important predictors of children's television viewing habits. Children who spend more time watching television tend to have fathers who are heavy television viewers. Also, children who watch many violent programs tend to have parents who watch violent programs such as *The Rockford Files, Quincy, Charlie's Angels, Wonder Woman, The Incredible Hulk, Starsky and Hutch, Vegas,* and *Baretta.* Cartoons were not included in this category, but were analysed separately.

Obviously, parents' viewing habits serve as a model for "appropriate" television viewing for their children. In addition, when we asked the children how they would feel if television "disappeared from this planet tomorrow", the children who were most upset were those whose parents watched the most television.

Developing a curriculum for school children concerning television

Over the years children have been taught to read and understand a newspaper format – to recognise the masthead, an editorial, a book review, the sports section, and special features (American Newspaper Publishers Association Foundation, 1978). They have also been taught about television news through lessons shown on instructional television. In addition, newspapers and radio stations (for instance, Boston's WGBH) have presented material to supplement classroom instruction. Although there have been several isolated attempts (Anderson and Ploghoft, 1980; Idaho State Department of Education, 1978; Media Action Research Center, 1979) to develop curricula to teach children about the general nature of television, major efforts in this area have been fairly recent. Many were stimulated by PTA interaction with parents, educators, and network officials about the negative consequences of television on children's lives. A number of major grants for curricula development for specific age groups were awarded in 1979, including contracts awarded by the US Office of Education.

With funding from American Broadcasting Companies, the Yale Family Television Research and Consultation Center developed a series of eight lessons for use in elementary schools to teach children to understand television and to capitalise on their interest in the medium in conjunction with reading, writing, and discussion skills. The study was carried out in a school system near a large city (Singer, Zuckerman and Singer, 1980), replicated twice in the same state. After many improvements in the videotapes and teachers' guides had been made, it was tested on a larger sample in ten school districts in cities scattered throughout the United States. An additional curriculum was also developed for kindergarten, first and second grade children, and subsequently tested in one community. This was a refinement of the elementary school curriculum but geared to younger children and emphasising more nonverbal activities. The results of all of these studies are included in this chapter.

Objectives of the lessons were as follows:

1. To understand the different types of television programs, such as news, documentaries, variety, game shows, situation comedies, dramas, etc.

2. To understand that programs are created by writers, producers, directors and other personnel, and that they they utilise actors and actresses as well as scenery and props.

3. To understand how television works in terms of simple electronics.

4. To learn what aspects of a program are real, and how fantasy or pretend elements are created for programs or commercials through camera techniques and special effects.

5. To learn about the purpose and types of commercials, including public service or political announcements.

6. To understand how television influences feelings, ideas, self-concept, and identification.

7. To become aware of television as a source of information about

other people, countries, and occupations and to become aware of the ways in which stereotypes are presented.

8. To help children to be more critical of violence on television; to become aware that television rarely shows someone recovering from an act of violence or the aggressor being punished; to understand the distinction between verbal and physical aggression.

9. To encourage children to be aware of what they watch and how they can control their viewing habits, as well as how they can influence networks, producers, and local television stations.

10. To use these lessons within a language and arts framework so that children can gain experience in using correct grammar and spelling, writing letters, abstracting ideas, critical thinking, expressive language, oral discussion, and reading.

The first study was carried out in a school system near New Haven, Connecticut, with third, fourth, and fifth grade children; there were 134 children in the experimental group and 98 in the control group. The children were matched for IQ, reading scores (on a standardised reading test), ethnicity and socio-economic status. The children in this study had a mean IQ of 110 and were about one year above their respective grade levels in reading. Parents kept records of their children's television viewing habits for two separate periods, for two weeks, before the experiment began, and for two weeks after the experiment was concluded.

The experimental condition consisted of eight lessons, two per week, taught during a four-week period by classroom teachers. Videotapes including clips from current television shows as well as original material illustrating the concepts to be taught were shown during each lesson. In addition, special materials including vocabulary lists, questions for discussion, activity sheets for classroom and homework assignments, and reference lists for books relating to each topic were included in each forty-minute lesson. In addition each student received a booklet describing television production and functions of television personnel. Emphasis was on teacher–pupil interaction, language and writing skills motivated by the children's interest in television, and the critical analysis of videotaped material. Classroom teachers were trained in the use of the materials, and parent workshops were held to acquaint them with the objectives and materials in the curriculum.

Evaluating the effects of the curriculum

Pre- and posttests and follow-up tests included questions related to the eight topics taught: Introduction to Television, Reality and Fantasy on Television, Camera Effects and Special Effects, Commercials and the Television Business, Identification with Television Characters,

Stereotypes on Television, Violence and Aggression, and How Viewers Can Influence Television. Three months later, the children in the experimental and control groups were retested, and the children in the control groups were then taught the same lessons. Pretest and posttest responses were compared in order to determine the effects of the lessons on knowledge and attitudes towards television. The children in the control school were compared to those in the experimental school in order to control for any changes that occurred as a result of the testing experience, the children's increased maturity, and other potentially influential factors.

Results indicated that children in the experimental school showed a greater increase in knowledge about television than those in the control school. Differences were greatest between the two groups in the measures of knowledge and understanding special effects, commercials, and advertising. The children in the experimental school also learned more lesson-related-vocabulary words and showed more improvement in their ability to identify videotaped examples of camera effects and special effects than children in the control school.

When comparisons between the experimental and control groups were made again three months after the experimental group had been taught, the original control group (after the lessons were taught to them) now learned even more than the original experimental group had learned in the first phase of the study. The dramatic increase in gains made by this group could be attributed to several factors: the control teachers benefiting from the feedback from the teachers of the original experimental group; the staff's familiarity with the material and resulting ease with the machinery and mechanics of each lesson; the awareness by the teachers that the lessons could work and that they could fit into a school curriculum without disruption; and their excitement and their desire to measure up to the first group's success.

In order to determine if the knowledge gained from the curriculum could be generalised to another situation, children were tested on new material involving the area of special effects. Results demonstrated that the children could transfer their learning to this new situation, and indeed understood camera techniques and special effects when used in different shows than the ones in the lessons.

Teachers and children were interested and attentive to most of the content in the eight lessons. The children were especially intrigued by the explanations of "disappearing", "bionic jumps", and slow and fast motion. Not only were the mechanics of television exciting to the children, but lively discussions took place dealing with topics such as stereotypes, aggression, and commercials. For example, when one third grade teacher asked which characters on television they liked, many boys said they liked *The Hulk*. When questioned about this preference, they replied, "he's big", "strong", "powerful", and "he can do anything". But when the teacher asked if the Hulk was "happy", the class responded with such answers as "No, he has a demon inside of him", "It's not good to be out of control", and "He never knows when he'll be a different person".

The children made many comments about watching violent programs. Several children said that they imitated characters on *CHiPS* or *Baretta* and that that led to trouble in the house. Another child said that "people learn from action shows not to do certain things". The children had watched such action shows as *Starsky and Hutch, CHiPS, The Incredible Hulk,* and *Charlie's Angels.*

A follow-up study and extension to younger children

After the initial pilot work was completed, we then proceeded to refine our curriculum and subsequently tested it again using a new sample in a different town in Connecticut. We also designed the materials for use with kindergarten, first and second grade children. Our new sample consisted of 91 kindergarten, first and second grade children and 135 third- and fourth-graders. In this study, children were taught the lessons once a week rather than the twice-a-week plan we had used in the pilot study. Materials for the younger children, kindergarten through second grade, were modified to include more emphasis on concrete and visual processing. Substantially more play activities were included such as a puppet stage and puppets, use of a "toy train" for sorting out passengers who were realistic, animated, real people, etc. We also reduced the number of lessons from 8 to 6 for the younger subjects, and used about 3 minutes of videotape material per lesson instead of the 7–10 minutes used in the original curriculum.

The teachers were again trained in the use of materials and parent-workshops were held to explain our purposes and present samples of the lessons.

The subjects, kindergarten through fourth grade, were in the same school; they were predominantly white and middle-class. In this study, a pre- and posttest design was employed consisting of a Television Comprehension Test. In addition, teacher ratings were obtained on the kindergarten, first and second grade children's behavior in terms of aggression, cooperation, general peer interaction and imagination. Teachers also rated the younger children, kindergarten through second grade, on reading or reading readiness ability, as well as on their cognitive (general learning) ability.

Results for kindergarten though second grade indicated that at each grade level there was clear evidence of improvement in understanding the material covered in the curriculum. Particularly strong effects were found for the children's learning about camera techniques, the nature of editing, and the distinction between reality and fantasy characters. This was of particular interest to us, since we were able to demonstrate that even five-, six- and seven-year-olds could learn about some of the technical aspects of television. It was also important to help some of these youngsters clarify the reality–fantasy distinction so that television events would be less frightening. For example, one child had experienced numerous nightmares about *The Hulk*. Once he learned about special effects, props, make-up and camera "tricks", his mother reported a decrease in his anxiety. One wonders why this youngster was permitted to view such a program in the first place, but given the television habits of American children and the lack of parental supervision *vis-à-vis* television, our approach has been to teach children about television in order to help them control their viewing habits, and to better understand the television conventions in terms of content and structure.

According to developmental psychologists, especially Piagetian theorists, second-graders would be expected to have a clearer under- standing of what is real and what is not real. Thus our results are particularly impressive since we found that not only the kindergarten children and first-graders learned this distinction, but second-graders made significant gains as well.

Additional findings for the younger children were significant for their awareness of what comprises a commercial, the meanings of disclaimers ("batteries not included", "each sold separately", etc.), and the purpose of commercials. The second grade children made especially significant improvements in their understanding of the different types of television programs such as cartoons, news and drama.

Overall, the third and fourth grade children made sizeable gains in understanding the different television categories, the awareness of edit- ing, evidence of generalisation from classroom examples of special effects to understanding effects used in a new situation not taught before. These older children also improved significantly on their understanding of special effects. Only "slow-motion" did not reflect a significant change in post-test scores, chiefly because most of the third- and fourth-graders identified this correctly *before* the lessons were taught.

We included in the pre- and posttest some questions that tapped the children's attitudes towards television, to see if exposure to a television curriculum might have any effect on these attitudes. For example, we were interested to know if there would be a shift in favorite television characters or programs. While there were trends towards a drop in the tendency to choose a violent character as their favorite, there were no significant shifts on many of the attitudinal questions. There was a highly significant result towards shifting the basis of judging a character as being a favorite from an emphasis on physical heroism towards other attributes. Third- and fourth-graders showed an increase in the tendency to choose a realistic rather than a fantasy figure from television as their favorite. When asked why they watched television, a strong tendency was found indicating they shifted from watching because programs were "exciting" towards judgments like "fun", or "nothing else to do" which may reflect somewhat less reliance on television for extreme stimulation.

There was a significant increase in the tendency for these older children to rely on newspapers rather than television for fuller information about the news. They were asked too, as in the pilot, how they would feel if television were to disappear from this planet. Prior to the lessons, third- and fourth-graders gave ratings indicating anger and sadness. Following the lessons, there was a decided trend for both grades for less extreme negative reaction to the possible loss of television.

Finally, we examined the relationship between teacher-rating variables and gain scores. The intercorrelations failed to reveal any meaningful trends. Although there are a number of significant correlations between gains on the Television Comprehension Test and teacher rating variables (in particular, children rated high on imagination gained more in such areas as audio-video distinction, fantasy aspects of television characters and props), the pattern of correlations is not consistent. Thus, we concluded that children benefit from the curriculum regardless of the level of their intellectual ability, reading or social skills. The course seems to work equally well for all the children.

National field-testing of the curriculum

The last phase of our curriculum study consisted of a national field testing of the final version of our videotapes and teacher–student manual. With additional funding from American Broadcasting Company, we were able to remake the videotapes in a more professional manner. We were also able to consolidate teacher lesson plans and students activities into one workbook, parts of which could be duplicated as the teacher deemed appropriate for the students. These new versions of the curriculum were now tested in ten school districts in different states. The sample of twelve classes consisted of 116 third-graders, 64 fourth-graders, 171 fifth-graders and 43 sixth-graders. In addition a third grade class took the course without videotapes and those data were analysed separately. There was one other group consisting of 19 talented and gifted fifth-graders whose data were also analysed separately.

Children were pre- and posttested on a Television Comprehension and Attitudes Test, and teachers assessed the curriculum using an Evaluation Form piloted in Connecticut. Unlike the previous testing in Connecticut, this field study was undertaken to evaluate the effectiveness of such a course with *no direct teacher training* by our staff. In addition we wanted to assess the changes and modifications in the videotapes.

Teachers in each participating school taught the lessons, one a week over an eight week period and were given four months lead time before the study began in order to become familiar with the manuals. They had about one month's time to review the videotapes before the lessons began in the Fall of 1980.

Results indicate that children were watching television on the average of 23.8 hours weekly for boys and 23.5 hours weekly for girls. Third-graders were the heaviest viewers averaging 26 hours per week. Following posttesting there were no significant decreases found in changes in amount of viewing. There was an increased tendency to examine television guides for programming and a shift away from more purely action-adventure shows toward exploration of public television and towards greater general selectivity.

Test items were organised into subgroups reflecting areas of coverage in order to analyse the data more cohesively. These included Production, Reality or Fantasy Distinctions, Commercials, Aggression, Genre or Types of Programming, Control of Television, and Stereotypes and Vocabulary items. On an overall basis across all schools, there were improvements in test scores for the 224 children in the nine schools in seven of these eight categories. Only the testing of issues relating to the administrative control of television failed to reflect improvement. Gains were strongest and especially statistically reliable for the categories of Production, Stereotypes and Vocabulary. Some schools showed gains in all areas of testing, while only one school failed to show gains in most areas. The teacher in this school seemed least prepared and was "less enthusiastic" about teaching about television in the classroom.

With regard to the "gifted" class, these students started out with higher pretest scores than the other children, thus there was less opportunity for gain scores when the pre- and posttest means were compared. However, both this class and the "no videotape class" made gains on awareness of television production items, reality and fantasy distinctions, understanding of commercials and television-related vocabulary. It was important to find that even without the use of videotapes, the teachers' manual and students' workbook were effective in conveying lesson content.

With respect to grade differences, the largest consistent gains emerged for the younger children, perhaps because they showed lower initial levels of comprehension of the various aspects of television. Viewed by grade and school, the heaviest gains were reflected in categories of Awareness of Production features, Awareness of Stereotyping, Commercials, Vocabulary and Reality and Fantasy distinctions. The problems of aggression or television self-control may be more difficult to communicate, or that more emphasis must be placed on these topics than the teachers did in this field test.

Analysis of the intercorrelations between the gains made from pre- to posttesting and factors such as amount of television-viewing, sex and grade, established that significant negative correlations existed between grade level and size of gains on practically all items. This suggests that younger children profited most from the lessons, at least as measured by testing. These gains were in all areas, suggesting that the lessons, designed as they originally were for the third- to fifth-graders, produced greatest gains on *all* categories for the younger children. Sex differences were generally negligible. The amount of television-viewing by children showed inverse relationships (although of small magnitude) to size of gains from lessons. We may assume that a third variable, such as IQ, achievement–orientation or some other correlate of heavy viewing may be playing a role in slightly limiting the progress made by heavy television-viewers in response to the lessons.

Twelve of the fifteen teachers who participated in the field study returned the Evaluation Forms. Ten of the teachers were "enthusiastic" and only two had reservations – one was "disappointed" and a third grade teacher said the lessons were "too hard". In terms of children's interest level, teachers described their interest as "high", and that they "enjoyed the videotapes". Overall response to the lessons was favorable, and the teachers rated the lesson on Aggression as "excellent", and the one on "Stereotypes" was generally a "favorite". Teachers agreed that the lessons "fit into a language arts and media skills" curriculum, and also agreed on the "need for such a program" because "children spend so much time watching TV".

It appears from our extensive testing of the television curriculum, that even in so brief time as 4–8 weeks, it is possible to introduce complicated subject matter on the nature of television and to demonstrate that children will learn the material. The lessons proved to be interesting to teach and were received for the most part positively by students. Pre-testing made it clear that elementary school children are ignorant about many aspects of the medium, television, although they watch it about three to four hours a day. It is clear that children enjoy learning about the "inside" world of television, are eager to participate in classroom discussions, and can utilise their newly acquired sophistication about television in their classroom and homework activities. Teachers were generally pleased with the materials and welcomed the opportunity to discuss television critically. They found that the workbooks tapped children's language skills such as vocabulary, sentence structure, abstracting ideas, letter writing, forming analogies and metaphors, and critical thinking.

It is up to the school systems to find a way of fitting such a course within the existing school structure. Based upon our experiences, we feel the television lessons could lend themselves to a more flexible way of presenting them within limited classroom time. One lesson could be taught over several sessions, or even during one term. Each lesson lends itself to a more extensive treatment and could be coordinated with other classroom topics. For example, a lesson on the "news" could be linked to the current events part of a class lesson or to social studies topics. A lesson on "Commercials" could blend into arithmetic or consumer affairs. As long as television plays such an important part in the daily lives of children, a constructive approach is to learn how to understand and "direct" it rather than let this mechanical box control us.

References

American Newspaper Publishers Association Foundation (1978). Teaching with news-papers: A newsletter for undergraduate method instructors. Washington, D.C.

Anderson, J. A. and Ploghoft, M. D. (1980). "Receivership Skills: The Television Experi-ence". Paper presented at the meeting of the International Communication Association, Acapulco, Mexico

Beuf, A. (1974). Doctor, lawyer, household drudge. *Journal of Communication* 24(2), 142–145

Blumenthal, A. L. (1977). *The Process of Cognition*. Prentice-Hall, Englewood Cliffs, N.J.

Blumler, J. G. and Katz, E. (1974). *The Uses of Mass Communications: Current Perspectives on Gratifications Research*. Sage, Beverly Hills, California

Busby, L. J. (1975). Sex-role research on the mass media. *Journal of Communication* 25(4), 35–44

Cantor, M. S. (1978). Where are the women in public broadcasting: *In* G. Tuchman, A. K. Daniels and J. Benet (eds) *Hearth and Home: Images of Women in the Mass Media*. Oxford University Press, New York

Collins, W. A. (1975). The developing child as a viewer. *Journal of Communication* 25(4), 35–44

Collins, W. A. (1978). Temporal integration and children's understanding of social informa-tion on television. *American Journal of Orthopsychiatry* 48, 198–204

Comstock, G., Chaffee, S., Katzman, N., McCombs, M. and Roberts, D. (1978). *Television and Human Behavior*. Columbia University Press, New York

Dembo, R. (1973). Gratifications found in media in British teenaged boys. *Journalism Quarterly* 6, 3

Dembo, R. and McCron, R. (1976). Social factors in media use. *In* R. Brown (ed.) *Children and Television*. Sage, Beverly Hills, California

Eron, L. D. (1980). Prescription for reduction of aggression. *American Psychologist* 35(3), 244–252

Frueh, T. and McGhee, P. E. (1975). Traditional sex role development and amount of time spent watching television. *Developmental Psychology* 11(1), 109

Furu, T. (1971). *The Function of Television for Children and Adolescents*. Sophia University Press, Tokyo

Greenberg, B. S. (1974). Gratifications of television viewing and the correlates for British children. *In* J. G. Blumler and E. Katz (eds) *The Uses of Mass Communications: Current Perspectives on Gratifications Research*. Sage, Beverly Hills, California

Greenberg, B. S. (1976). Viewing and listening parameters among British youngsters. *In* R. Brown (ed.) *Children and Television*. Sage, Beverly Hills, California

Greenberg, B. S. (1982). Television and role socialization. *In* D. Pearl, L. Bouthilet and J. Lazar (eds) *Television and Behavior: Ten Years of Scientific Progress and Implications for the Eighties*. US Government Printing Office, Washington, D.C.

Greenberg, B. S. and Dominick, J. R. (1970). Television behavior among disadvantaged children. *In* B. S. Greenberg and B. Dervin (eds) *Use of the Mass Media by Urban Poor*. Praeger, New York

Greenstein, J. (1954). Effects of television upon elementary school grades. *Journal of Educational Research* 48(3), 161–176

Hornik, R. (1978). Television access and the slowing of cognitive growth. *American Educational Research Journal* 15, 1–15

Idaho State Department of Education (1978). "The Way We See It: A Program to Improve Critical Television Skills". Boise

Lefkowitz, M. M., Eron, L. D., Walder, L. O. and Huesmann, L. R. (1977). *Growing up to be Violent*. Pergamon Press, Elmsford, New York

Lemar, J. (1977). Women and blacks on prime-time television. *Journal of Communication* 27, 70–80

Liebert, R. M. and Schwartzberg, N. S. (1977). Effects of mass media. In *Annual Review of Psychology*, Vol. 28. Annual Reviews, Palo Alto

Lyle, J. and Hoffman, H. R. (1972). Children's use of television and other media. *In* E. A. Rubinstein, G. A. Comstock and J. P. Murray (eds) *Television and Social Behavior*, Vol. 4,

Television in Day-to-Day Life: Patterns of Use. US Government Printing Office, Washington, D.C.

Media Action Research Center (1979). *Television Awareness Training.* New York

Meyer, T. P. (1976). Impact of "All in the Family" on children. *Journal of Broadcasting* **20**(1), 23–33

Murray, J. P. and Kippax, S. (1978). Children's social behavior in three towns with differing television experience. *Journal of Communication* **28**(1), 19–29

O'Bryant, S. L. and Corder-Bolz, C. R. (1978). The effects of television on children's stereotyping of women's work roles. *Journal of Vocational Behavior* **12**, 233–244

Postman, N. (1980). Fine-tuning the balance between education and a media culture. *Teacher* **98**, 28–30

Potter, R. L. (1976). *New Season: The Positive Use of Commercial Television with Children.* Charles E. Merrill, Columbus, Ohio

Potter, R. L., Faith, C. and Ganek, L. B. (1979). *Channel. Critical Reading/TV Viewing Skills.* Educational Activities Inc., Freeport, New York

Potter, R. (1981). "Making the Most of Television". Weekly column, *St. Petersburg Times,* Florida

Ridder, J. (1963). Public Opinion and the relationship of TV viewing to academic achievement. *Journal of Educational Research* **57**(4), 204–207

Roper, B. W. (1980). "Evolving Public Attitudes Towards Television and other Mass Media 1959–1980". A report by the Roper Organization, Inc. Television Information Office, New York

Schramm, W., Lyle, J. and Parker, E. (1961). *Television in the Lives of Our Children.* Stanford University Press, Stanford

Singer, D. G. (1982). Television and the developing imagination of the child. *In* D. Pearl, L. Bouthilet and J. Lazar (eds) *Television and Behavior: Ten Years of Scientific Progress and Implications for the Eighties.* US Government Printing Office, Washington, D.C.

Singer, D. G. and Singer, J. L. (1981–1982). "Television and the family". Monthly column in *TV Guide*

Singer, D. G. and Stoving, R. L. *Television tune-ins.* Goodyear Publishing Company Inc., Santa Monica, California (In press)

Singer, D. G., Singer, J. L. and Dodsworth-Rugani, K. J. (1979). "'Fables of the Green Forest' and 'Swiss Family Robinson': An Experimental Evaluation of their Educational and Prosocial Potential". Unpublished manuscript, Yale University

Singer, D. G., Zuckerman, D. M. and Singer, J. L. (1980). Helping elementary school children learn about TV. *Journal of Communication* **30**(3), 84–93

Singer, D. G., Singer, J. L. and Zuckerman, D. M. (1981). *Getting the most out of TV.* Goodyear Publishing Company Inc., Santa Monica, California

Singer, J. L. (1979). "Television and Reading in the Development of Imagination". Paper presented at the conference sponsored by Deutsche Lesegesellschaft, Mainz, West Germany

Singer, J. L. and Singer, D. G. (1981). *Television, Imagination and Aggression: A Study of Preschoolers.* Erlbaum, Hillsdale, N.J.

Surgeon General's Scientific Advisory Committee on Television and Social Behavior. (1972). *Television and Growing up: The Impact of Televised Violence.* US Government Printing Office, Washington, D.C.

Thompson, C. (1964). Children's acceptance of television advertising and the relation of televiewing to school achievement. *Journal of Educational Research* **58**(4), 171–175

Tower, R. B., Singer, D. G., Singer, J. L. and Biggs, A. (1979). Differential effects of television programming on preschoolers' cognition, imagination, and social play. *American Journal of Orthopsychiatry* **49**, 265–281

von Feilitzen, C. (1976). The functions served by the media: Report on a Swedish study. *In* R. Brown (ed.) *Children and Television.* Sage, Beverly Hills, California

Witelson, S. F. (1976). Sex and the single hemisphere: Specialization of the right-hemisphere for spatial processing. *Science* **193**, 425–427

Witty, P. (1967). Children of the TV era. *Elementary English* **44**, 528–535

Zuckerman, D. M., Singer, D. G. and Singer, J. L. (1980a). Children's television viewing, racial and sex role attitudes. *Journal of Applied Psychology* **10**(4), 281–294

Zuckerman, D. M., Singer, D. G. and Singer, J. L. (1980b). Television viewing and children's reading and related classroom behavior. *Journal of Communication* **30**(1), 166–174

Subject Index